PRACTICE - ASSESS - DIAGNO

180 Days of GEOGRAPHY for Fourth Grade

Author
Chuck Aracich, M.A.Ed.

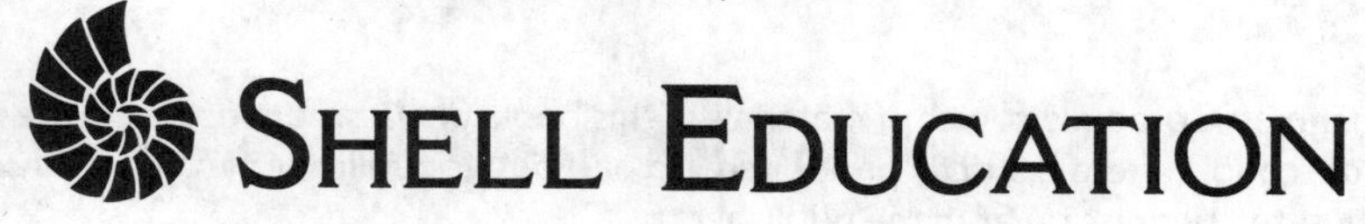

Series Consultant

Nicholas Baker, Ed.D.
Supervisor of Curriculum and Instruction
Colonial School District, DE

Publishing Credits

Corinne Burton, M.A.Ed., *Publisher*
Conni Medina, M.A.Ed., *Managing Editor*
Emily R. Smith, M.A.Ed., *Content Director*
Veronique Bos, *Creative Director*
Shaun N. Bernadou, *Art Director*
Lynette Ordoñez, *Editor*
Kevin Pham, *Graphic Designer*
Stephanie Bernard, *Associate Editor*

Image Credits

p.108 Eileen 10/Shutterstock; p.138 (bottom left and center) Jonathan Weiss/Shutterstock; p.138 (bottom right) Jessica Kirsh/Shutterstock; p.148 Omar C. Garcia; all other images from iStock and/or Shutterstock

Standards

For information on how this resource meets national and other state standards, see pages 10–14. You may also review this information by visiting our website at www.teachercreatedmaterials.com/administrators/correlations/ and following the on-screen directions.

Shell Education
A division of Teacher Created Materials
5301 Oceanus Drive
Huntington Beach, CA 92649-1030
www.tcmpub.com/shell-education

ISBN 978-1-4258-3305-3

TABLE OF CONTENTS

INTRODUCTION

With today's geographic technology, the world seems smaller than ever. Satellites can accurately measure the distance between any two points on the planet and give detailed instructions about how to get there in real time. This may lead some people to wonder why we still study geography.

While technology is helpful, it isn't always accurate. We may need to find detours around construction, use a trail map, outsmart our technology, and even be the creators of the next navigational technology.

But geography is also the study of cultures and how people interact with the physical world. People change the environment, and the environment affects how people live. People divide the land for a variety of reasons. Yet no matter how it is divided or why, people are at the heart of these decisions. To be responsible and civically engaged, students must learn to think in geographical terms.

The Need for Practice

To be successful in geography, students must understand how the physical world affects humanity. They must not only master map skills but also learn how to look at the world through a geographical lens. Through repeated practice, students will learn how a variety of factors affect the world in which they live.

Understanding Assessment

In addition to providing opportunities for frequent practice, teachers must be able to assess students' geographical understandings. This allows teachers to adequately address students' misconceptions, build on their current understandings, and challenge them appropriately. Assessment is a long-term process that involves careful analysis of student responses from a discussion, project, practice sheet, or test. The data gathered from assessments should be used to inform instruction: slow down, speed up, or reteach. This type of assessment is called *formative assessment.*

HOW TO USE THIS BOOK

Weekly Structure

The first two weeks of the book focus on map skills. By introducing these skills early in the year, students will have a strong foundation on which to build throughout the year. Each of the remaining 34 weeks will follow a regular weekly structure.

Each week, students will study a grade-level geography topic and a state. Some weeks will focus on two states to ensure that all 50 can be covered in one school year.

Days 1 and 2 of each week focus on map skills. Days 3 and 4 allow students to apply information and data to what they have learned. Day 5 helps students connect what they have learned to themselves.

Day 1—Reading Maps: Students will study a grade-appropriate map and answer questions about it.

Day 2—Creating Maps: Students will create maps or add to an existing map.

Day 3—Read About It: Students will read a text related to the topic or location for the week and answer text-dependent or photo-dependent questions about it.

Day 4—Think About It: Students will analyze a chart, diagram, or other graphic related to the topic or location for the week and answer questions about it.

Day 5—Geography and Me: Students will do an activity to connect what they learned to themselves.

Five Themes of Geography

Good geography teaching encompasses all five themes of geography: location, place, human-environment interaction, movement, and region. Location refers to the absolute and relative locations of a specific point or place. The place theme refers to the physical and human characteristics of a place. Human-environment interaction describes how humans affect their surroundings and how the environment affects the people who live there. Movement describes how and why people, goods, and ideas move between different places. The region theme examines how places are grouped into different regions. Regions can be divided based on a variety of factors, including physical characteristics, cultures, weather, and political factors.

HOW TO USE THIS BOOK *(cont.)*

Weekly Themes

The following chart shows the topics, states, and themes of geography that are covered during each week of instruction.

	Topic	State(s)	Geography Themes
1	—Map Skills Only—		Location
2			Location
3	Fishing Industry	Massachusetts/Maine	Place, Human-Environment Interaction
4	Beaches	New Jersey	Location, Human-Environment Interaction
5	Pollution	New York	Place, Human-Environment Interaction
6	Climate	Rhode Island/ Connecticut	Place
7	Natural Resources	Pennsylvania	Location, Place, Human-Environment Interaction
8	Watersheds	New Hampshire/ Vermont	Place, Human-Environment Interaction, Region
9	Forests	Virginia/West Virginia	Location, Place
10	Counties	Georgia	Region
11	Rivers	Alabama/Mississippi	Human-Environment Interaction, Movement
12	Everglades	Florida	Location, Place, Region
13	Oceans	Maryland/Delaware	Location, Place, Human-Environment Interaction
14	Renewable/ Nonrenewable Resources	Arkansas	Human-Environment Interaction
15	Hurricanes	North Carolina/South Carolina	Human-Environment Interaction
16	Swamps/Bayous	Louisiana	Place, Human-Environment Interaction
17	Plateaus	Kentucky/Tennessee	Place, Region
18	Population/Cities	Illinois	Place, Human-Environment Interaction
19	Tourism	Ohio/Indiana	Place, Movement

HOW TO USE THIS BOOK *(cont.)*

	Topic	State(s)	Geography Themes
20	Points of Interest	North Dakota/South Dakota	Place
21	Erosion	Missouri	Human-Environment Interaction, Region
22	Land Use	Nebraska/Kansas	Place, Human-Environment Interaction
23	Agriculture	Wisconsin/Iowa	Place
24	Great Lakes	Michigan/Minnesota	Place, Human-Environment Interaction
25	Tornados	Oklahoma	Human-Environment Interaction, Region
26	Desert	Arizona/New Mexico	Place, Human-Environment Interaction, Region
27	Cultural Landscapes	Texas	Place
28	Westward Expansion	Utah	Place, Movement
29	Deforestation and Reforestation	Washington/Oregon	Human-Environment Interaction
30	Mountains	Colorado	Location, Place, Region
31	Trade	Idaho	Movement
32	The Great Plains	Wyoming/Montana	Region
33	Droughts	Nevada	Place, Human-Environment Interaction
34	Earthquakes	California	Human-Environment Interaction
35	Volcanoes	Hawai'i	Place
36	Tundra	Alaska	Human-Environment Interaction, Region

HOW TO USE THIS BOOK *(cont.)*

Using the Practice Pages

The activity pages provide practice and assessment opportunities for each day of the school year. Teachers may wish to prepare packets of weekly practice pages for the classroom or for homework.

As outlined on page 4, each week examines one location and one geography topic.

The first two days focus on map skills. On Day 1, students will study a map and answer questions about it. On Day 2, they will add to or create a map.

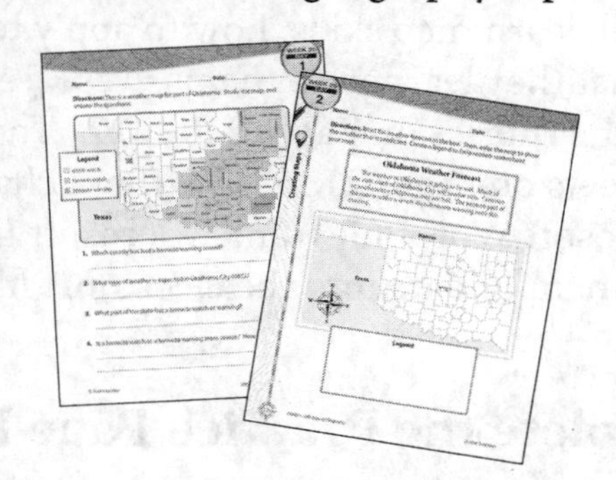

Days 3 and 4 allow students to apply information and data from texts, charts, graphs, and other sources to the location being studied.

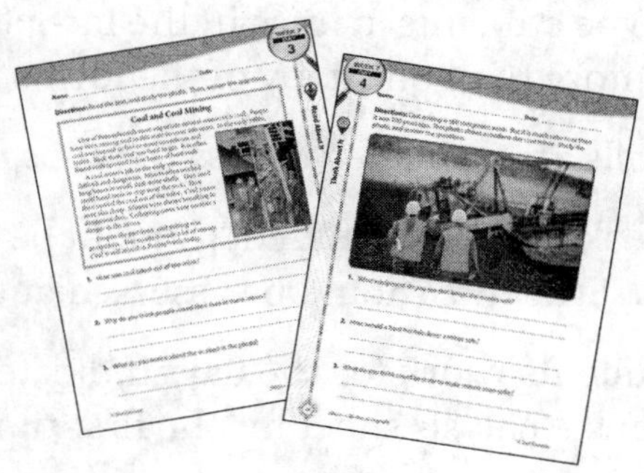

On Day 5, students will apply what they learned to themselves.

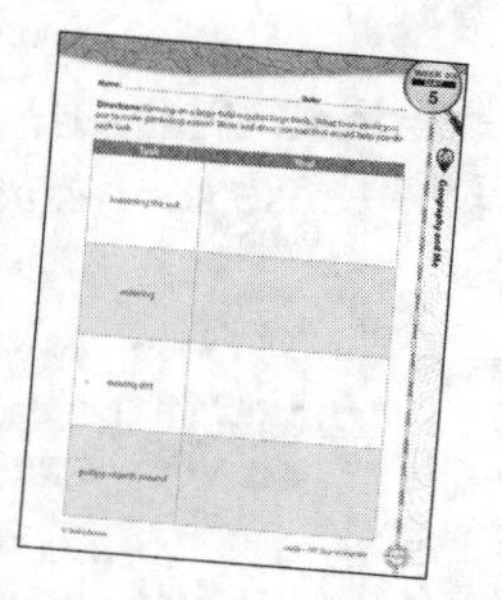

Using the Resources

Rubrics for the types of days (map skills, applying information and data, and making connections) can be found on pages 210–212 and in the Digital Resources. Use the rubrics to assess students' work. Be sure to share these rubrics with students often so that they know what is expected of them.

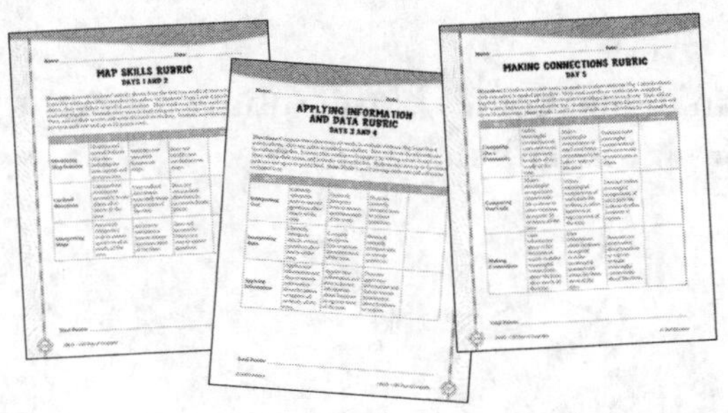

HOW TO USE THIS BOOK *(cont.)*

Diagnostic Assessment

Teachers can use the practice pages as diagnostic assessments. The data analysis tools included with the book enable teachers or parents to quickly score students' work and monitor their progress. Teachers and parents can quickly see which skills students may need to target further to develop proficiency.

Students will learn map skills, how to apply text and data to what they have learned, and how to relate what they learned to themselves. Teachers can assess students' learning in each area using the rubrics on pages 210–212. Then, record their scores on the Practice Page Item Analysis sheets on pages 213–215. These charts are also provided in the Digital Resources as PDFs, Microsoft Word® files, and Microsoft Excel® files (see page 216 for more information). Teachers can input data into the electronic files directly on the computer, or they can print the pages.

To Complete the Practice Page Item Analyses:

- Write or type students' names in the far-left column. Depending on the number of students, more than one copy of the forms may be needed.
    - The skills are indicated across the tops of the pages.
    - The weeks in which students should be assessed are indicated in the first rows of the charts. Students should be assessed at the ends of those weeks.
- Review students' work for the days indicated in the chart. For example, if using the Making Connections Analysis sheet for the first time, review students' work from Day 5 for all five weeks.
- Add the scores for each student. Place that sum in the far right column. Record the class average in the last row. Use these scores as benchmarks to determine how students are performing.

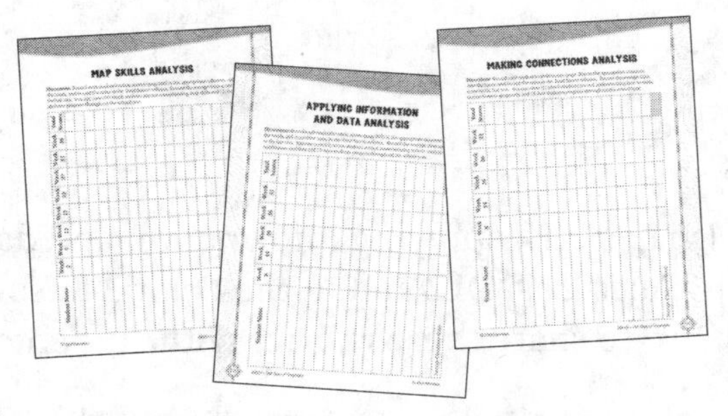

Digital Resources

The Digital Resources contain digital copies of the rubrics, analysis pages, and standards charts. See page 216 for more information.

Using the Results to Differentiate Instruction

Once results are gathered and analyzed, teachers can use them to inform the way they differentiate instruction. The data can help determine which geography skills are the most difficult for students and which students need additional instructional support and continued practice.

Whole-Class Support

The results of the diagnostic analysis may show that the entire class is struggling with certain geography skills. If these concepts have been taught in the past, this indicates that further instruction or reteaching is necessary. If these concepts have not been taught in the past, this data is a great preassessment and may demonstrate that students do not have a working knowledge of the concepts. Thus, careful planning for the length of the unit(s) or lesson(s) must be considered, and additional front-loading may be required.

Small-Group or Individual Support

The results of the diagnostic analysis may show that an individual student or a small group of students is struggling with certain geography skills. If these concepts have been taught in the past, this indicates that further instruction or reteaching is necessary. Consider pulling these students aside to instruct them further on the concepts while others are working independently. Students may also benefit from extra practice using games or computer-based resources.

Teachers can also use the results to help identify proficient individual students or groups of students who are ready for enrichment or above-grade-level instruction. These students may benefit from independent learning contracts or more challenging activities.

STANDARDS CORRELATIONS

Shell Education is committed to producing educational materials that are research and standards based. In this effort, we have correlated all our products to the academic standards of all 50 states, the District of Columbia, the Department of Defense Dependents Schools, and all Canadian provinces.

How to Find Standards Correlations

To print a customized correlation report of this product for your state, visit our website at **www.teachercreatedmaterials.com/administrators/correlations** and follow the on-screen directions. If you require assistance in printing correlation reports, please contact our Customer Service Department at 1-877-777-3450.

Purpose and Intent of Standards

The Every Student Succeeds Act (ESSA) mandates that all states adopt challenging academic standards that help students meet the goal of college and career readiness. While many states already adopted academic standards prior to ESSA, the act continues to hold states accountable for detailed and comprehensive standards. Standards are designed to focus instruction and guide adoption of curricula. Standards are statements that describe the criteria necessary for students to meet specific academic goals. They define the knowledge, skills, and content students should acquire at each level. Standards are also used to develop standardized tests to evaluate students' academic progress. Teachers are required to demonstrate how their lessons meet state standards. State standards are used in the development of our products, so educators can be assured they meet the academic requirements of each state.

The activities in this book are aligned to the National Geography Standards and the McREL standards. The chart on pages 11–12 lists the National Geography Standards used throughout this book. The chart on pages 13–14 correlates the specific McREL and National Geography Standards to each week. The standards charts are also in the Digital Resources (standards.pdf).

C3 Framework

This book also correlates to the College, Career, and Civic Life (C3) Framework published by the National Council for the Social Studies. By completing the activities in this book, students will learn to answer and develop strong questions (Dimension 1), critically think like a geographer (Dimension 2), and effectively choose and use geography resources (Dimension 3). Many activities also encourage students to take informed action within their communities (Dimension 4).

STANDARDS CORRELATIONS *(cont.)*

180 Days of Geography is designed to give students daily practice in geography through engaging activities. Students will learn map skills, how to apply information and data to their understandings of various locations and cultures, and how to apply what they learned to themselves.

Easy to Use and Standards Based

There are 18 National Geography Standards, which fall under six essential elements. Specific expectations are given for fourth grade, eighth grade, and twelfth grade. For this book, fourth grade expectations were used.

Essential Elements	National Geography Standards
The World in Spatial Terms	**Standard 1:** How to use maps and other geographic representations, geospatial technologies, and spatial thinking to understand and communicate information
	Standard 2: How to use mental maps to organize information about people, places, and environments in a spatial context
	Standard 3: How to analyze the spatial organization of people, places, and environments on Earth's surface
Places and Regions	**Standard 4:** The physical and human characteristics of places
	Standard 5: People create regions to interpret Earth's complexity
	Standard 6: How culture and experience influence people's perceptions of places and regions
Physical Systems	**Standard 7:** The physical processes that shape the patterns of Earth's surface
	Standard 8: The characteristics and spatial distribution of ecosystems and biomes on Earth's surface

STANDARDS CORRELATIONS *(cont.)*

Essential Elements	National Geography Standards
Human Systems	**Standard 9:** The characteristics, distribution, and migration of human populations on Earth's surface
	Standard 10: The characteristics, distribution, and complexity of Earth's cultural mosaics
	Standard 11: The patterns and networks of economic interdependence on Earth's surface
	Standard 12: The process, patterns, and functions of human settlement
	Standard 13: How the forces of cooperation and conflict among people influence the division and control of Earth's surface
Environment and Society	**Standard 14:** How human actions modify the physical environment
	Standard 15: How physical systems affect human systems
	Standard 16: The changes that occur in the meaning, use, distribution, and importance of resources
The Uses of Geography	**Standard 17:** How to apply geography to interpret the past
	Standard 18: How to apply geography to interpret the present and plan for the future

–2012 National Council for Geographic Education

STANDARDS CORRELATIONS *(cont.)*

Easy to Use and Standards Based *(cont.)*

This chart lists the specific National Geography Standards and McREL standards that are covered each week.

Wk.	NGS	McREL Standards
1	Standards 1 and 3	Knows the basic elements of maps and globes. Knows major physical and human features of places as they are represented on maps and globes.
2	Standards 1 and 3	Knows the basic elements of maps and globes. Knows major physical and human features of places as they are represented on maps and globes.
3	Standard 11	Knows economic activities that use natural resources in the local region, state, and nation and the importance of the activities to these areas.
4	Standard 4	Knows human-induced changes that are taking place in different regions and the possible future impacts of these changes.
5	Standard 4	Knows the ways in which the physical environment is stressed by human activities.
6	Standard 7	Understands how physical processes help to shape features and patterns on Earth's surface.
7	Standard 16	Knows economic activities that use natural resources in the local region, state, and nation and the importance of the activities to these areas.
8	Standards 14 and 18	Knows the ways in which the physical environment is stressed by human activities.
9	Standard 4	Knows how the characteristics of places are shaped by physical and human processes.
10	Standard 9	Knows the functions of political units and how they differ on the basis of scale.
11	Standard 12	Knows how transportation and communication have changed and how they have affected trade and economic activities.
12	Standard 8	Knows the components of ecosystems at a variety of scales.
13	Standard 4	Knows the physical components of Earth's atmosphere, lithosphere, hydrosphere, and biosphere.
14	Standard 16	Knows the characteristics, location, and use of renewable resources, flow resources, and nonrenewable resources.
15	Standard 15	Knows natural hazards that occur in the physical environment.
16	Standard 15	Knows the physical components of Earth's atmosphere, lithosphere, hydrosphere, and biosphere.

STANDARDS CORRELATIONS *(cont.)*

Wk.	NGS	McREL Standards
17	Standards 4 and 12	Knows the physical components of Earth's atmosphere, lithosphere, hydrosphere, and biosphere.
18	Standards 12 and 14	Knows the characteristics and locations of cities and how cities have changed over time.
19	Standard 12	Understands ways in which people view and relate to places and regions differently.
20	Standard 17	Knows how the characteristics of places are shaped by physical and human processes.
21	Standard 7	Understands how physical processes help to shape features and patterns on Earth's surface.
22	Standards 12 and 14	Knows the ways people alter the physical environment.
23	Standard 14	Knows how human activities have increased the ability of the physical environment to support human life in the local community, state, United States, and other countries.
24	Standard 15	Knows how communities benefit from the physical environment.
25	Standard 15	Knows natural hazards that occur in the physical environment.
26	Standard 8	Knows plants and animals associated with various vegetation and climatic regions on Earth.
27	Standards 10 and 17	Understands cultural change.
28	Standard 17	Knows the geographic factors that have influenced people and events in the past.
29	Standard 16	Knows the ways people alter the physical environment.
30	Standard 7	Knows the physical components of Earth's atmosphere, lithosphere, hydrosphere, and biosphere.
31	Standard 11	Knows the various ways in which people satisfy their basic needs and wants through the production of goods and services in different regions of the world.
32	Standard 5	Knows the characteristics of a variety of regions.
33	Standard 15	Knows the ways in which human activities are constrained by the physical environment. Knows natural hazards that occur in the physical environment.
34	Standard 15	Knows natural hazards that occur in the physical environment.
35	Standard 4	Knows the physical components of Earth's atmosphere, lithosphere, hydrosphere, and biosphere.
36	Standard 8	Knows the characteristics of a variety of regions.

Name: ______________________________ Date: ______________

Directions: Follow the steps to complete the map. Use the symbols in the legend.

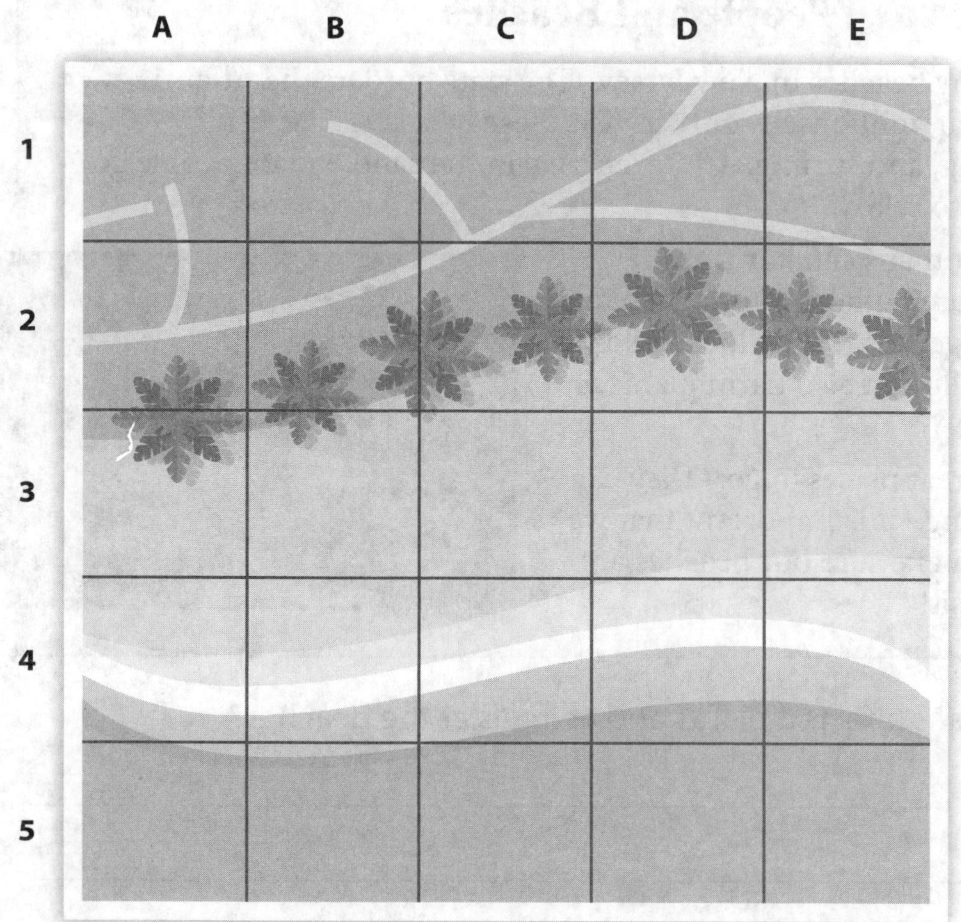

1. Draw lifeguard towers in D4 and A3.

2. Draw a trash can in B3.

3. Draw a first aid station in C1.

4. Draw a bathroom in D2.

5. Draw a parking lot in E1.

6. Draw a snack stand in E3.

Read About It

Name: ______________________________ **Date:** ________________

Directions: Read the text, and study the photo. Then, answer the questions.

People and Beaches

There are many beaches in New Jersey. The state borders the Atlantic Ocean. Every year, millions of tourists visit these beaches. They enjoy swimming, sailing, and surfing. Children play in the sand. Some people go to the beach just to relax.

Sometimes, humans can harm the beaches. Hotels are built on animal habitats. Litter can be dangerous for the local wildlife. Air and water pollution also harm plants and animals.

Beaches are great places to go. They can be fun and relaxing. It is important that we do all we can to make sure our beaches are great places to visit!

1. How have humans affected plants and animals at the beaches?

____

____

____

2. What do you see in the photo that might be harmful to a beach habitat?

____

____

3. What are some other beach activities not mentioned in the text?

____

____

____

Name: ______________________________ **Date:** ______________

Directions: This graph shows the number of people who visit New Jersey and its beaches each year. Study the graph, and answer the questions.

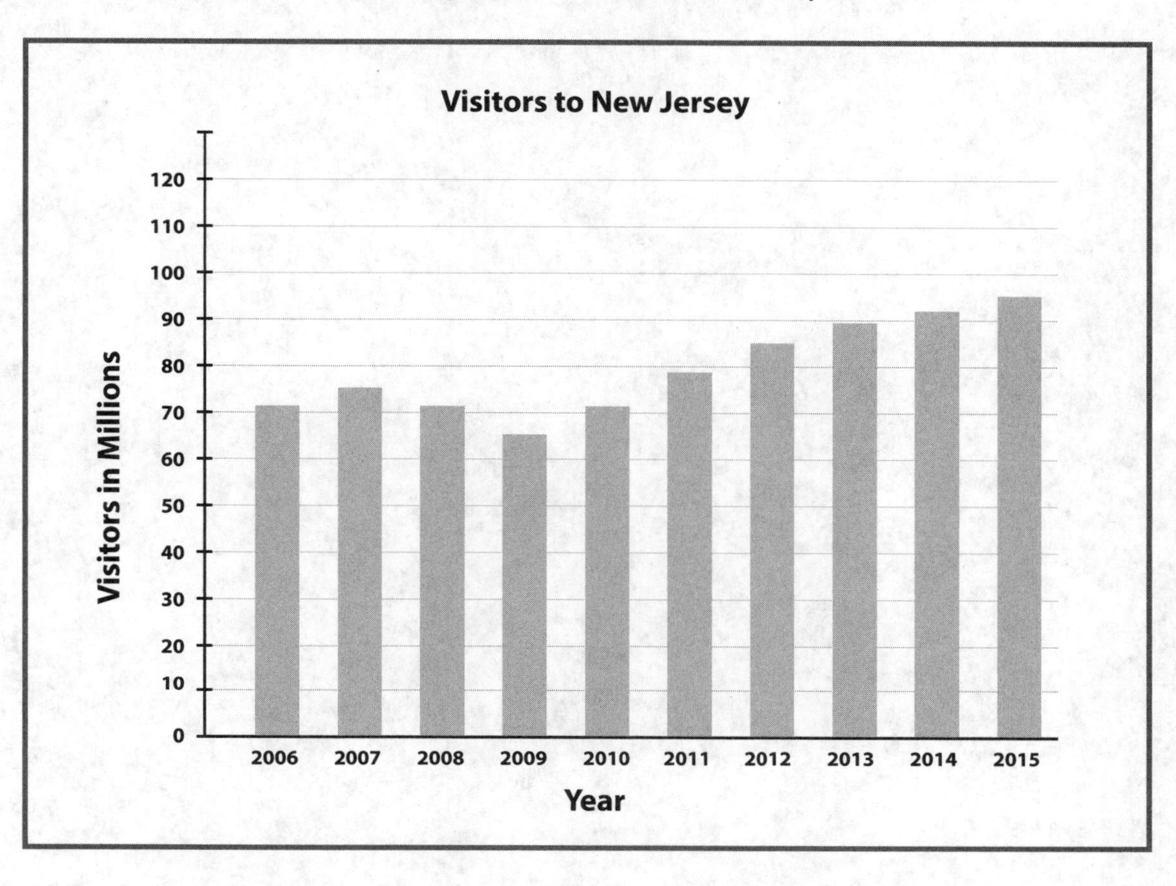

1. About how many more people visited New Jersey in 2015 than in 2009?

____

2. Are more or fewer people visiting New Jersey over time? How do you know?

____

____

____

____

3. How many years are represented on this graph?

____

Name: ______________________________ Date: ________________

Directions: Create a sign reminding people to be responsible and protect the beaches.

Name: ______________________ **Date:** ______________

Directions: This map shows trash and recycling centers in a town. Study the map, and answer the questions.

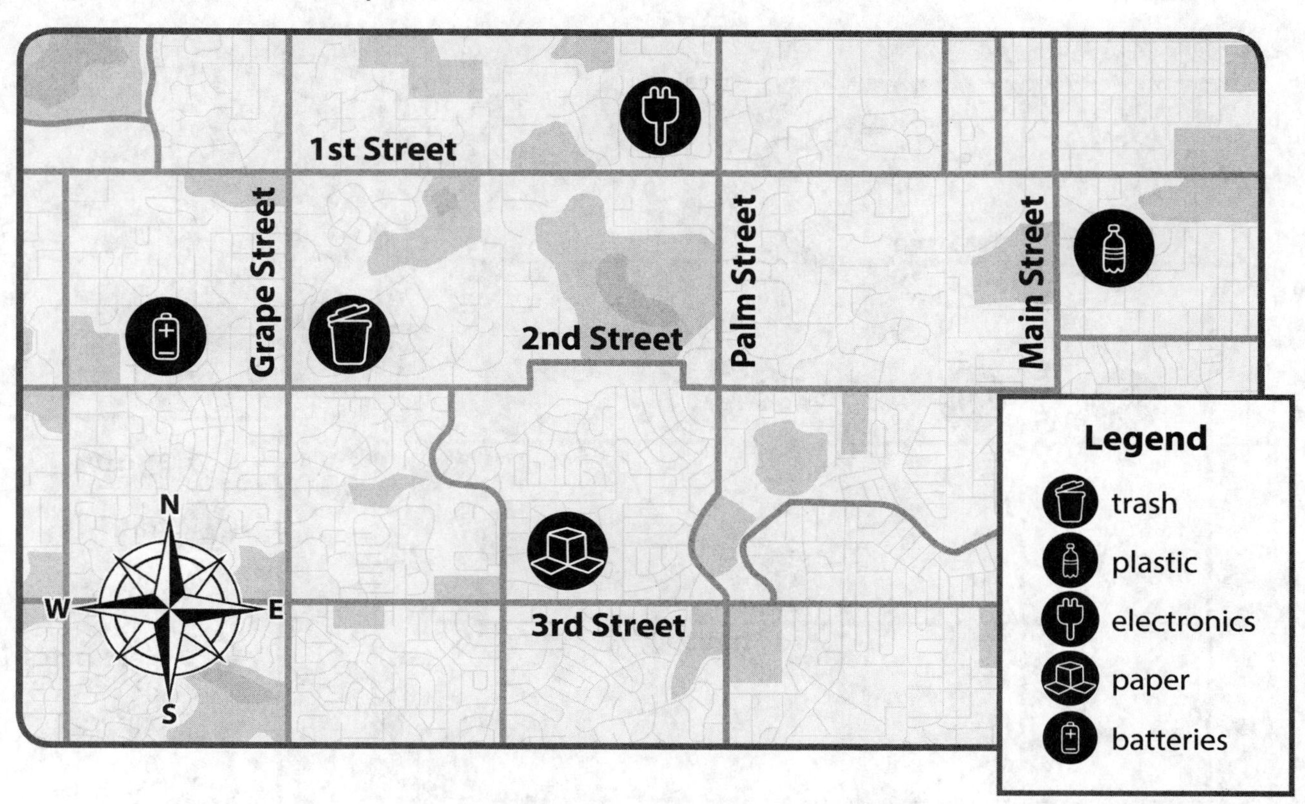

1. Which items can be recycled on 1st Street?

____

2. Give directions to go from the plastic recycling area to the battery disposal area. Include street names and directions.

____

____

____

3. Why might it be better to have all recycling and disposal areas in one location?

____

____

Creating Maps

Name: ______________________ Date: ______________

Directions: A composting center is going to be built in the town. Follow the steps, and answer the question.

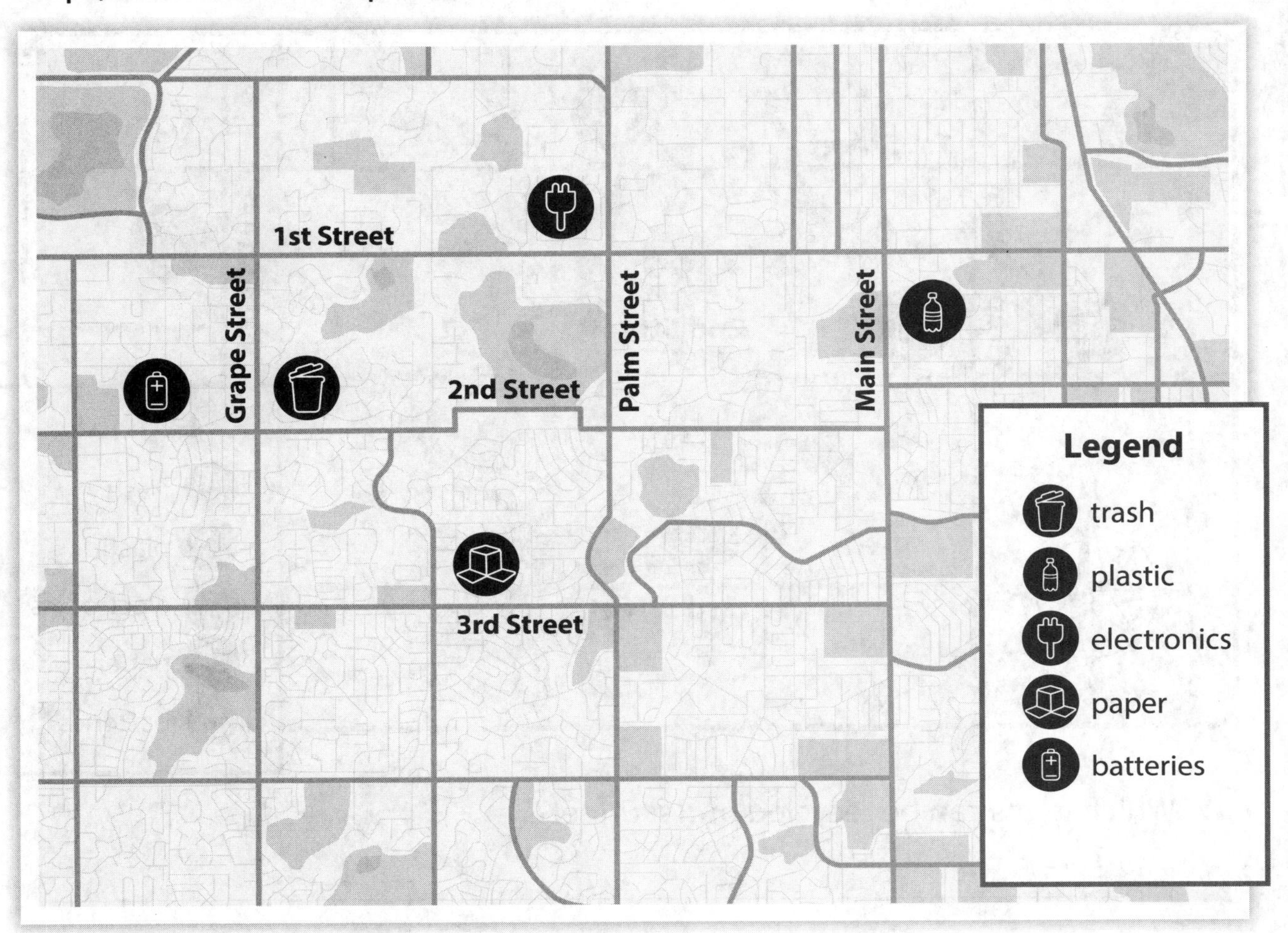

1. Choose a location for the composting center. Draw a symbol for the composting center in the location you chose.

2. Add your symbol to the legend.

3. Why is this a good location for the composting center?

____

____

____

____

Name: ______________________ Date: ____________

Directions: Read the text, and study the photo. Then, answer the questions.

Different Types of Pollution

All major cities have pollution. New York City is no exception. Millions of people live and travel its busy streets. A city with this many people easily creates a lot of pollution.

One of the most common types of pollution is air pollution. Smoke and chemicals from vehicles get into the air. Businesses and factories also release gases into the air. This makes it hard for people to breathe. Water pollution is also a problem. Trash and waste from the city make their way to nearby lakes and rivers.

Some pollution is less common. City life can be noisy. Cars, trucks, and planes create a lot of noise. This is a form of noise pollution. Noise pollution can cause health issues for people. Light pollution can be harmful, too. Billboards, homes, and buildings produce light. This can make nighttime city streets nearly as bright as in the daytime. These nighttime lights disrupt wildlife. Sometimes, the lights can be seen for miles.

1. What are some of the more uncommon types of pollution?

____

2. Name two types of noise pollution that may be taking place in the photo.

____

____

3. How might people in the photo reduce pollution?

____

____

Think About It

Name: ______________________ **Date:** ______________

Directions: Noise pollution is a major issue in New York City. This chart shows the number of noise complaints in New York City from 2013 to 2014. Study the chart, and answer the questions.

Complaint Type	Count
loud music/party	52,368
construction	28,999
loud talking	18,210
car/truck music	8,962
barking dog	7,480
air conditioning	4,200
car idling	3,886
car horn	3,374
banging/pounding	3,087
other	10,098

1. How many total complaints were caused by vehicles?

____

2. How many more complaints were there about barking dogs compared to car horns?

____

3. Is it possible to eliminate noise pollution? Explain your thinking.

____

____

____

____

Name: ______________________ Date: ______________

Directions: Sometimes, you can reuse items instead of throwing them away. How might you reuse these items? Give an example for each.

Plastic Grocery Bag

Cardboard Box

Gallon Milk Jug

Old T-Shirt

Reading Maps

Name: ____________________ Date: ____________

Directions: This map shows the average lowest temperatures in Connecticut and Rhode Island. Study the map, and answer the questions.

Average Low Temperatures

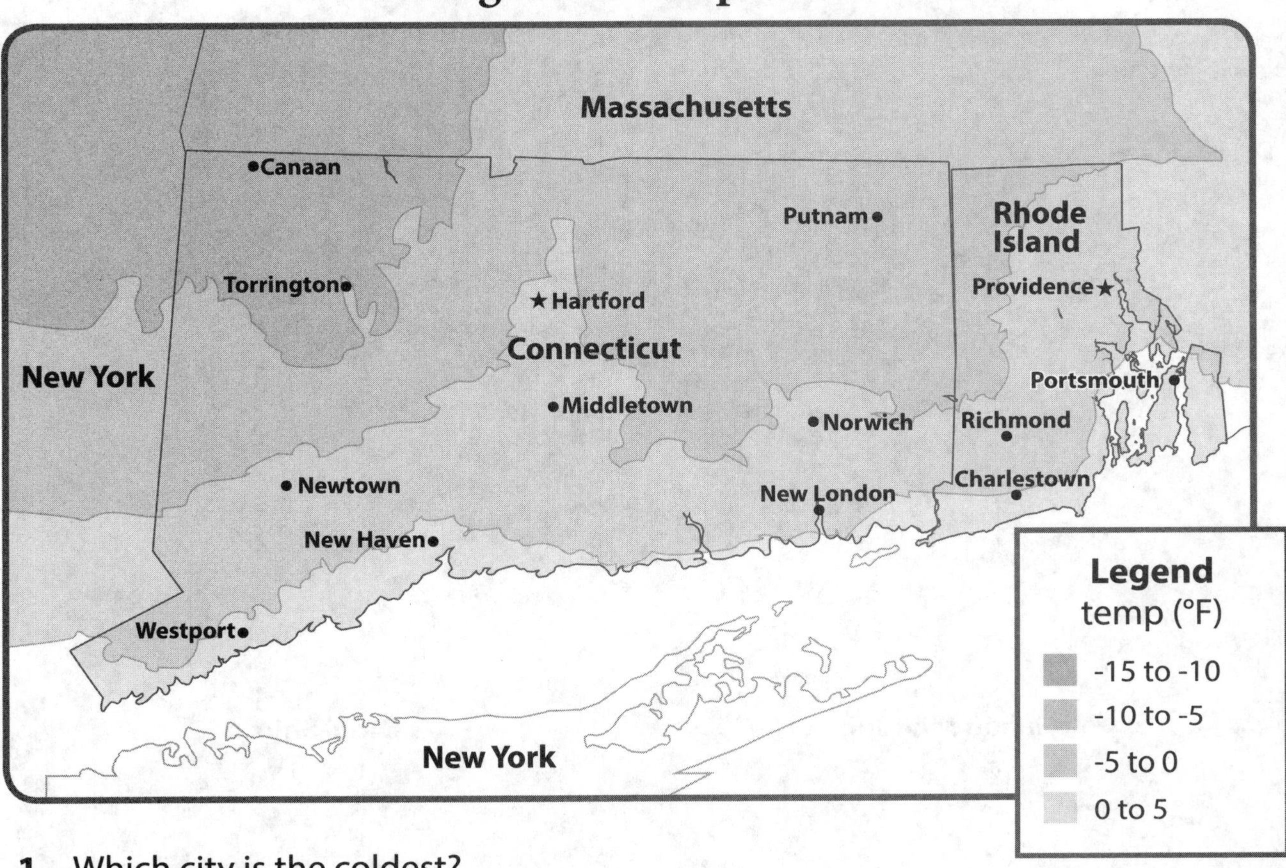

1. Which city is the coldest?

____

2. Which season is shown on this map? How do you know?

____

____

3. Based on this map, would you rather live in Connecticut or Rhode Island? Why?

____

____

Name: ______________________ Date: ______________

Directions: Follow the steps to complete the map.

____

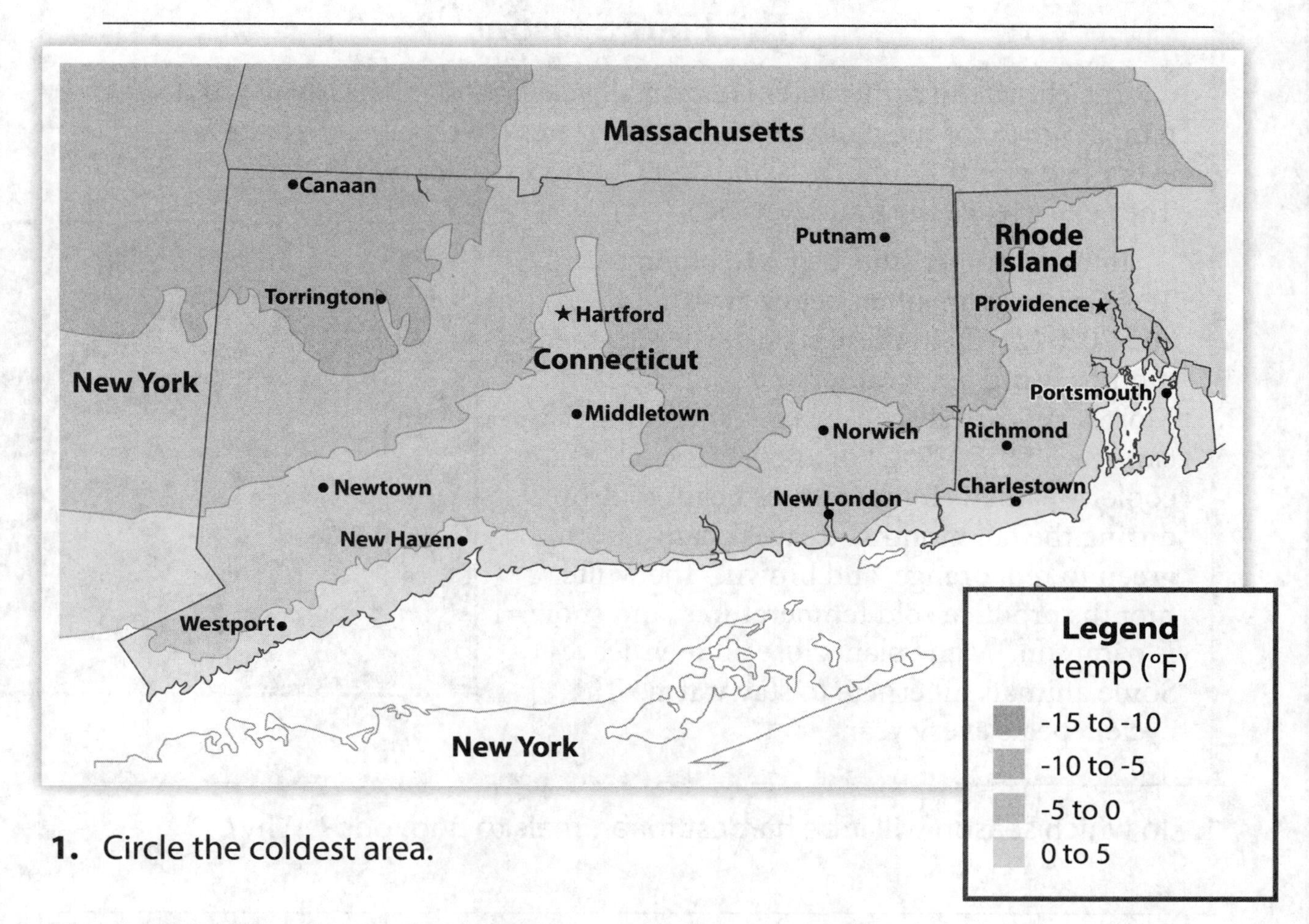

1. Circle the coldest area.

2. Lightly shade the warmest area.

3. Outline Connecticut in red. Outline Rhode Island in blue.

4. Put a box around each state's capital city.

5. Add a compass rose to the map.

6. Draw a triangle next to the city that is the farthest east in each state.

7. Add a title to the map.

Read About It

Name: ______________________ Date: ______________

Directions: Read the text, and study the photo. Then, answer the questions.

The Four Seasons

The climate in some states is warm all year round. Others have cold temperatures for most of the year. Many states have a climate that changes every few months. Rhode Island and Connecticut are two of these states. They experience the four seasons.

In the spring, plants begin to bloom. Temperatures are often between 40°F (4°C) and 70°F (21°C) in these states. The summer months bring hot weather and long days. Most plants are at their peak. Temperatures can reach 90°F (32°C) or above. The fall has cooler weather. Temperatures begin to drop during the fall months. Leaves change from green to red, orange, and brown. The winter months produce cold temperatures, and snow is common. Many plants lose all their leaves. Some animals hibernate to stay warm. This cycle repeats every year.

1. In which season will it be hardest for animals to find food? Why?

____

____

2. Describe what happens in the fall.

____

____

3. How might these seasonal changes affect people living in Connecticut and Rhode Island?

____

____

Name: ______________________________ **Date:** ______________

Directions: This line graph shows the hours of daylight in Rhode Island throughout the year. Use the graph to answer the questions.

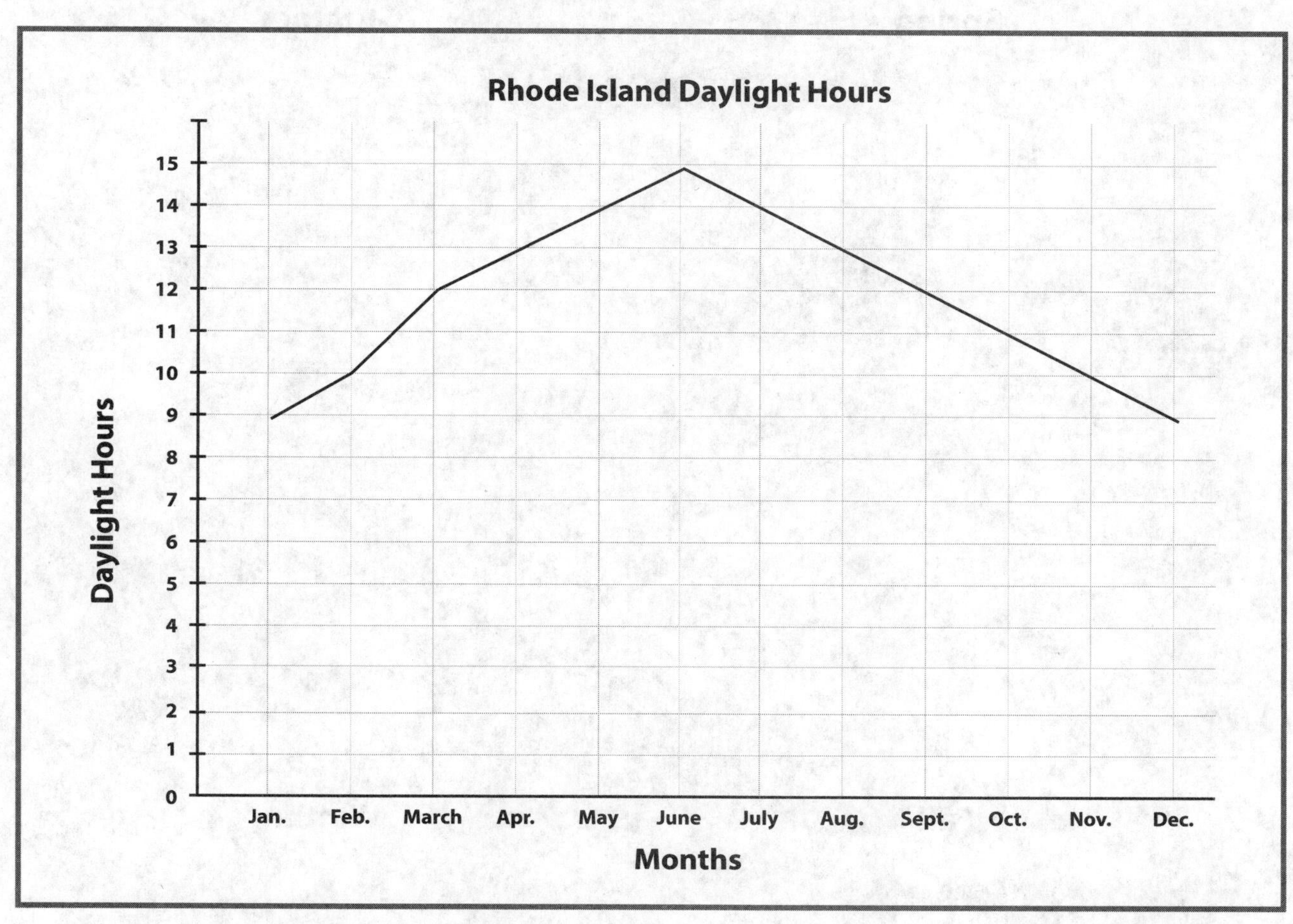

1. Which month has about the same amount of sunlight as September?

____

2. Which month do you think would be the warmest? Why do you think so?

____

____

3. Will this information change from year to year? Explain your thinking.

____

____

Name: ______________________ Date: ______________

Directions: Write or draw in each box to describe the climate where you live.

Name: ________________________________ Date: ________________

Directions: States can be divided into regions. The land in each region shares common characteristics. Study the map, and answer the questions.

Pennsylvania Regions

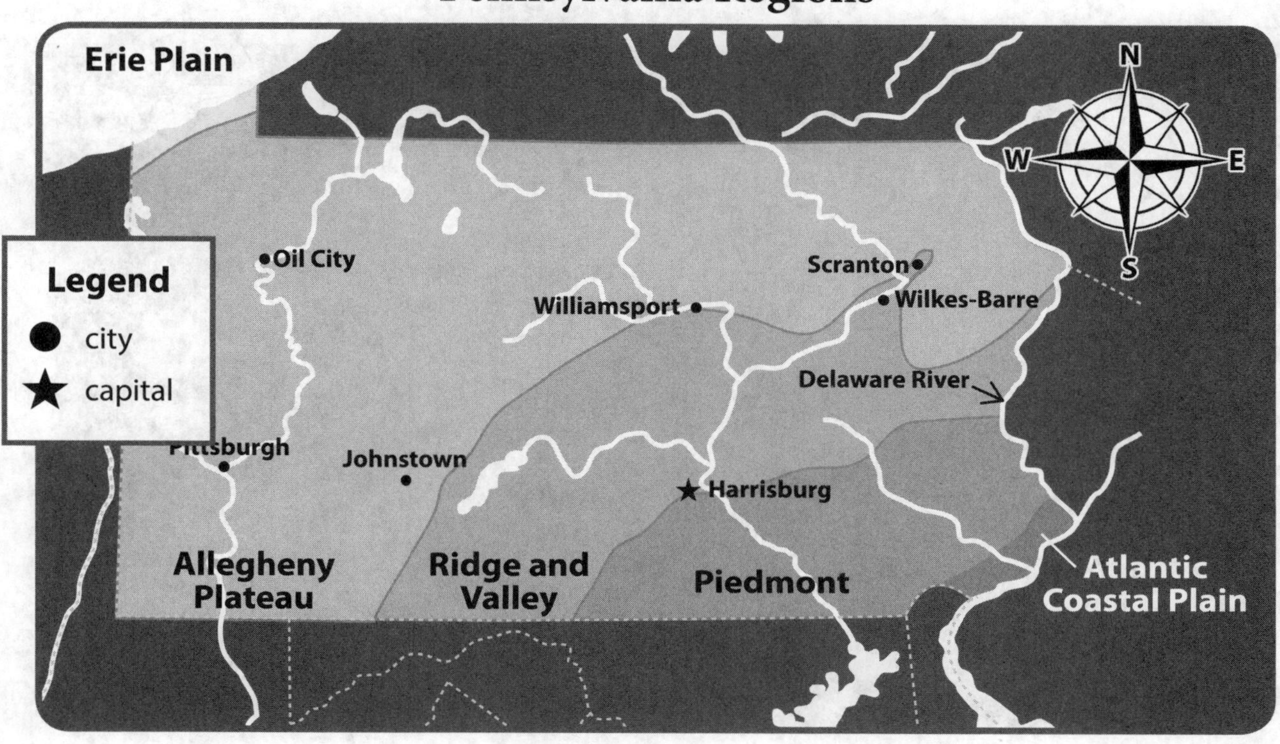

1. In which region is the capital of Pennsylvania? What symbol is used to represent the capital?

____

____

2. There are coal mines in northeastern Pennsylvania. Which region is this?

____

3. Which regions does the Delaware River border?

____

____

Name: ______________________ **Date:** ______________

Directions: Water is an important natural resource. Use the clues to label some of Pennsylvania's rivers and lakes.

Pennsylvania Regions

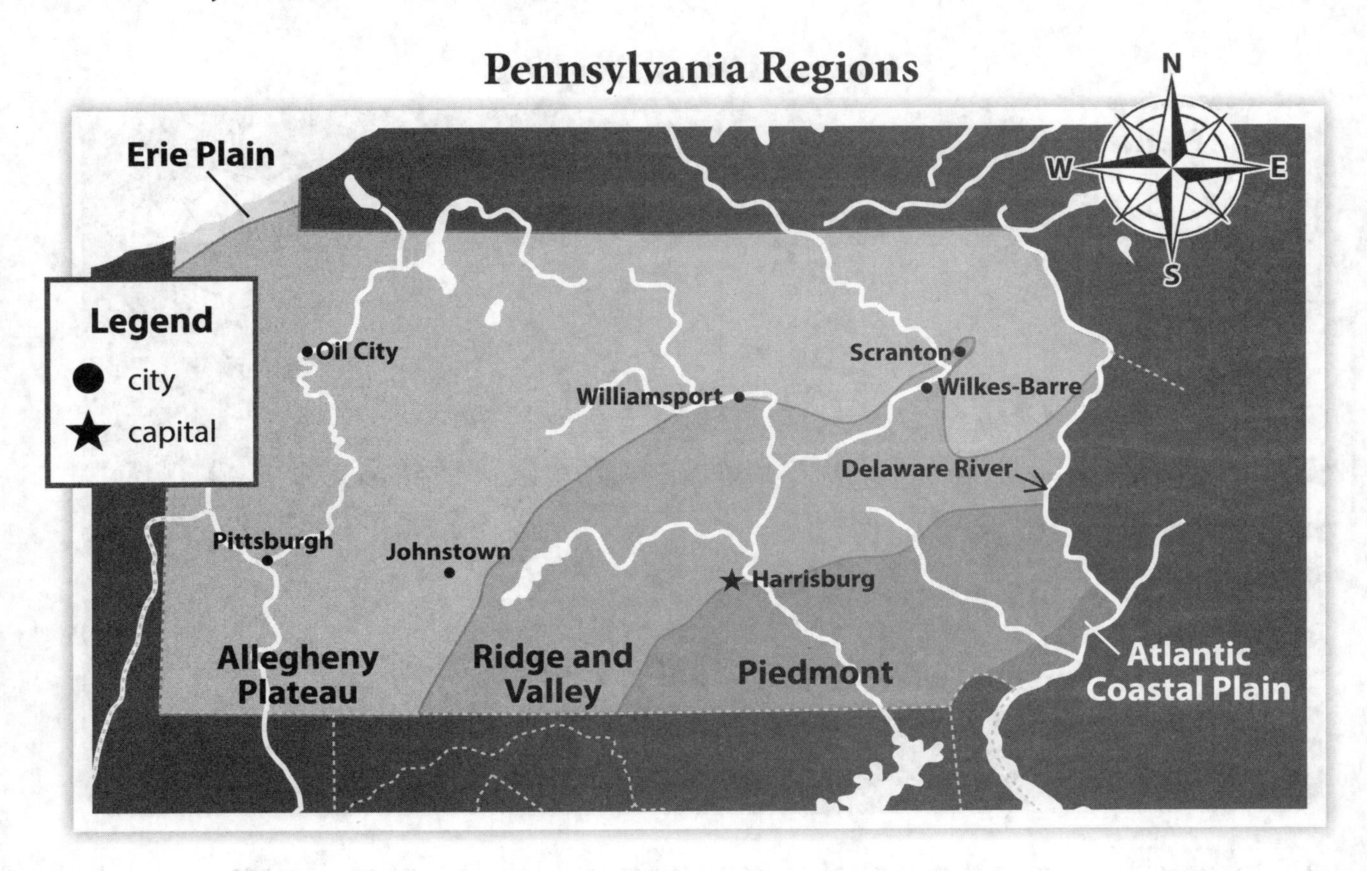

1. The Ohio River flows southeast near Pittsburgh in the Allegheny Plateau region.
2. The Allegheny River meets the Ohio River in Pittsburgh.
3. The Schuylkill River flows into the Delaware River in the Atlantic Coastal Plain region.
4. Lake Erie is located on Pennsylvania's northwest border.
5. The Susquehanna River extends from Pennsylvania's northern border to its southern border.
6. Raystown Lake is west of Harrisburg in the Ridge and Valley region.

Name: ______________________ **Date:** ______________

Directions: Read the text, and study the photo. Then, answer the questions.

Coal and Coal Mining

One of Pennsylvania's most important natural resources is coal. People have been mining coal in this state for over 100 years. In the early 1900s, coal was burned as fuel to move steamboats and trains. Back then, coal was hard to get. It is often found underground below layers of hard rock.

A coal miner's job in the early 1900s was difficult and dangerous. Miners often worked long hours in small, dark mine shafts. They used small hand tools to chip away the rock. They then carried the coal out of the mine. Coal mines were also dusty. Miners were always breathing in dangerous dust. Collapsing caves were another a danger in the mines.

Despite the problems, coal mining was prosperous. This means it made a lot of money. Coal is still mined in Pennsylvania.

1. How was coal taken out of the mine?

____

____

2. Why do you think people risked their lives in these mines?

____

____

3. What do you notice about the workers in the photo?

____

____

Think About It

Name: ______________________________ **Date:** ______________

Directions: Coal mining is still dangerous work. But it is much safer now than it was 100 years ago. This photo shows modern-day coal mines. Study the photo, and answer the questions.

1. What equipment do you see that keeps the miners safe?

____

____

2. How would a hard hat help keep a miner safe?

____

____

3. What do you think could be done to make miners even safer?

____

____

Name: ______________________ Date: ______________

Directions: Write how you benefit from each of these natural resources.

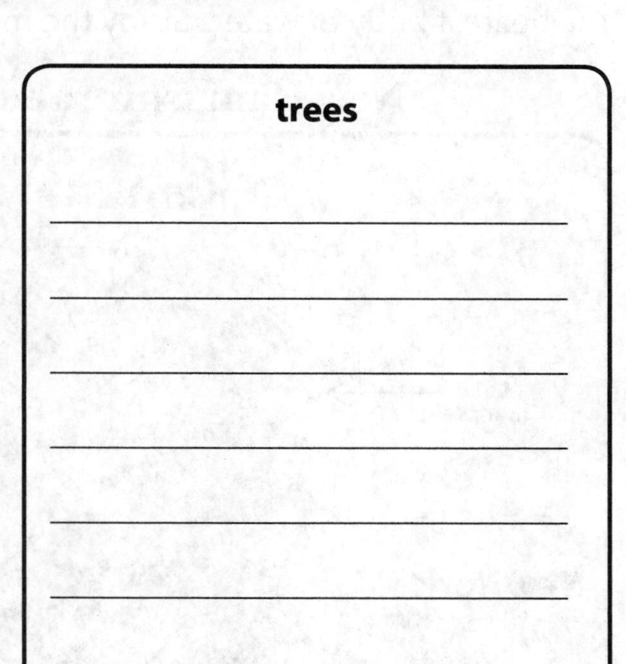

sunlight

trees

Natural Resources

water

soil

Name: ______________________ Date: ____________

Directions: A watershed is the area of land that captures water and drains into the nearest body of water. Study the map, and answer the questions.

New Hampshire and Vermont Watersheds

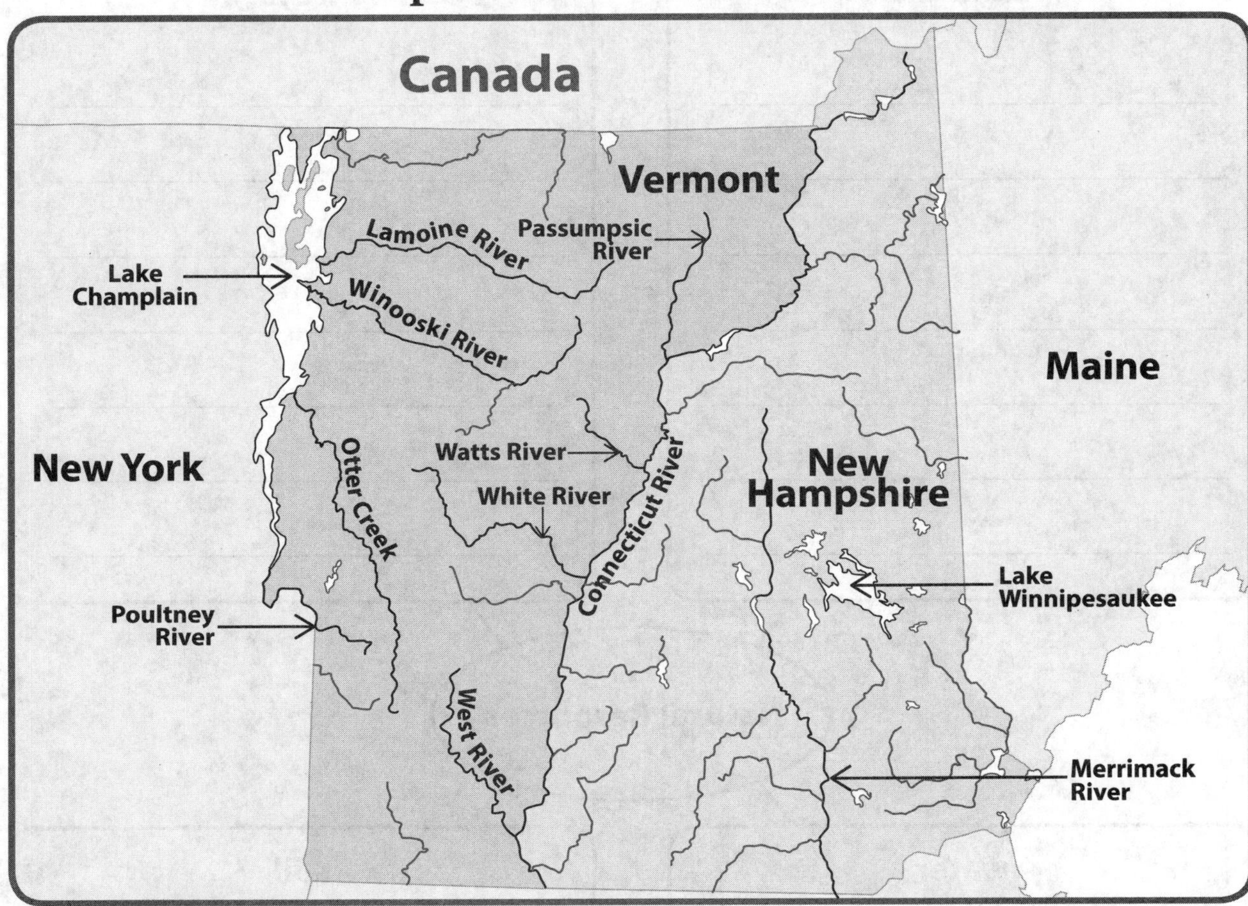

1. What are the names of the two lakes shown on this map?

____

____

2. Which river forms the border between New Hampshire and Vermont?

____

3. Name two rivers in Vermont that appear to drain into the river from question 2.

____

____

Name: ______________________________ **Date:** ________________

Directions: Follow the steps to complete the map.

New Hampshire and Vermont Watersheds

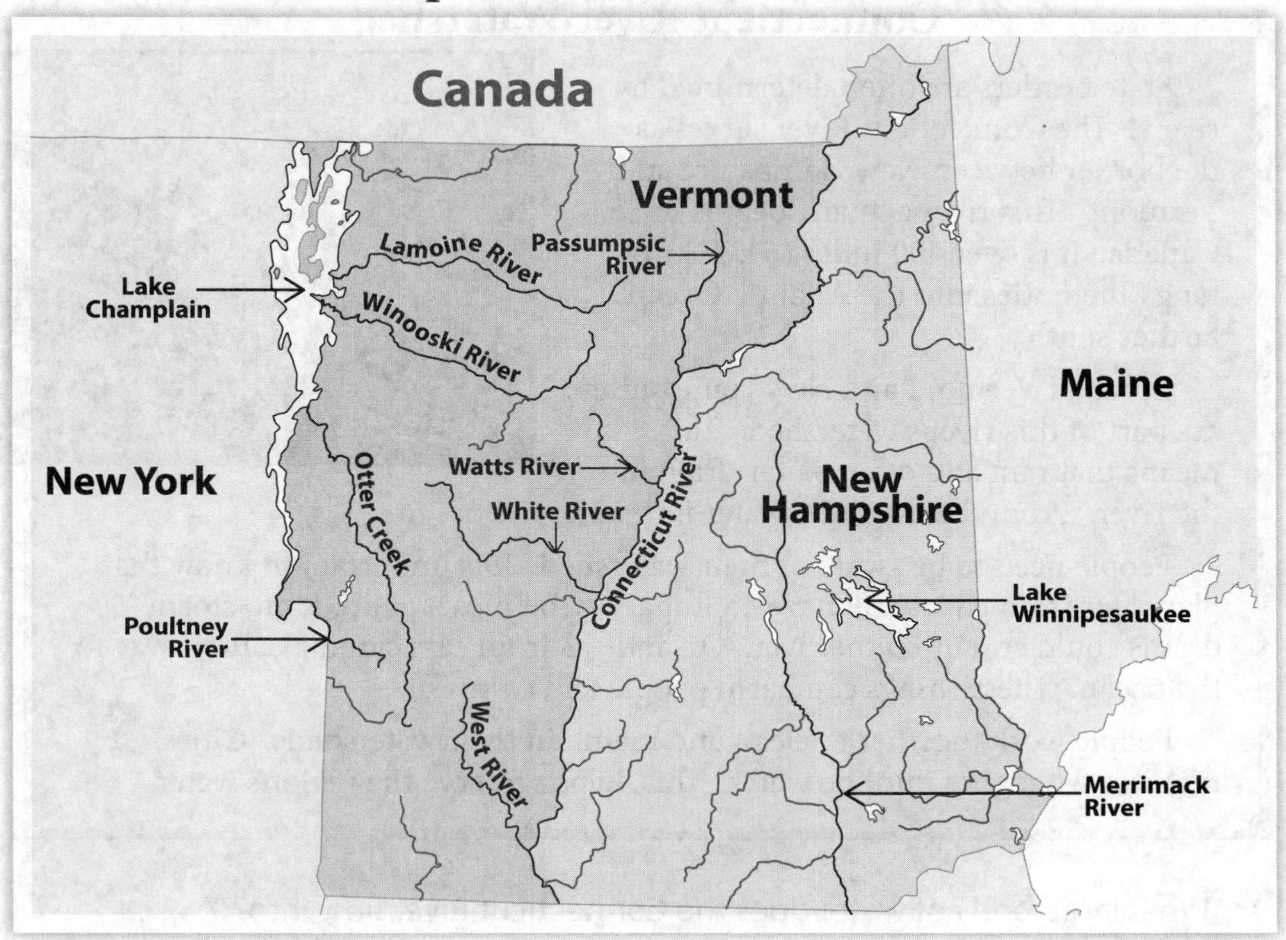

1. Trace the Connecticut River in blue.

2. Much of the water in Vermont and New Hampshire flows into the Connecticut River. Trace each river from both states that flows into the Connecticut River.

3. Many other rivers flow into the Merrimack River in New Hampshire. Use a different color to trace those rivers.

4. Many other rivers in Vermont flow into Lake Champlain. Use a third color to trace those rivers.

Name: ______________________ **Date:** ____________

Read About It

Directions: Read the text, and study the photo. Then, answer the questions.

Connecticut River Watershed

State borders are often determined by rivers. The Connecticut River serves as the border between New Hampshire and Vermont. This river actually begins in Canada. It is over 400 miles (643.7 km) long. It empties into the Atlantic Ocean farther south.

Much of Vermont and New Hampshire are part of this river's watershed. This means that rain and other water drain into the river. From the river, the water flows into the ocean.

People need to be aware of their watershed. It is important to know that places far from rivers still have an impact. Chemicals poured into storm drains could end up in the river. Oil and gas from cars can make their way to the ocean. These things can harm people and animals.

People work together to clean and maintain their watersheds. Cities monitor what goes into the water. This helps preserve the region's water.

1. What large body of water does the Connecticut River flow into?

____

2. Why is it important to keep watersheds clean?

____

____

3. Why might rivers make good state borders?

____

____

Name: ______________________________ **Date:** ______________

Directions: Study the photo, and read the facts about the Connecticut River cleanup. Then, answer the questions.

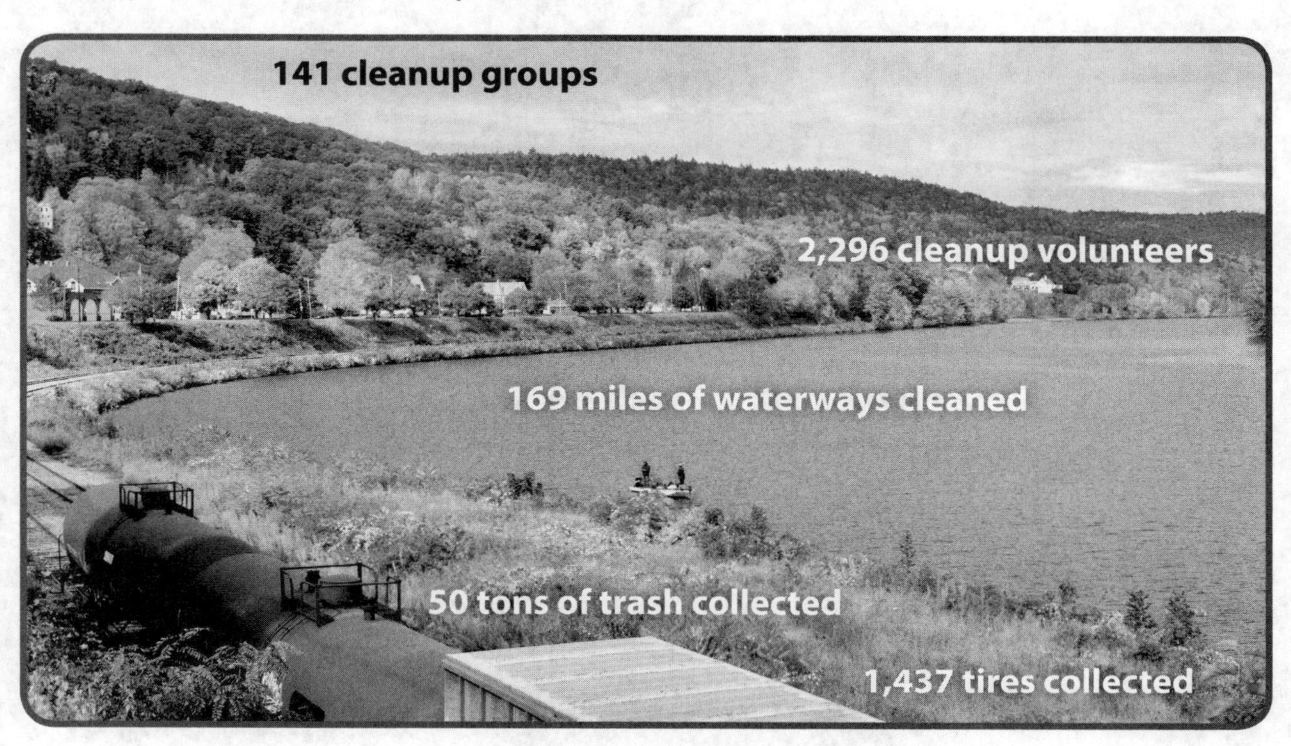

1. What might make cleaning a river difficult?

____

____

2. Do you think this photo best represents a cleanup? Why or why not?

____

____

____

3. Which fact surprised you most? Why?

____

____

Name: ______________________________ Date: ________________

Directions: Everyone can make a difference when it comes to watersheds! Draw and write to show what you could do to keep a watershed healthy.

____

____

____

____

____

____

____

____

Name: ______________________ **Date:** ______________

Directions: Study the map, and answer the questions.

Virginia State Forests

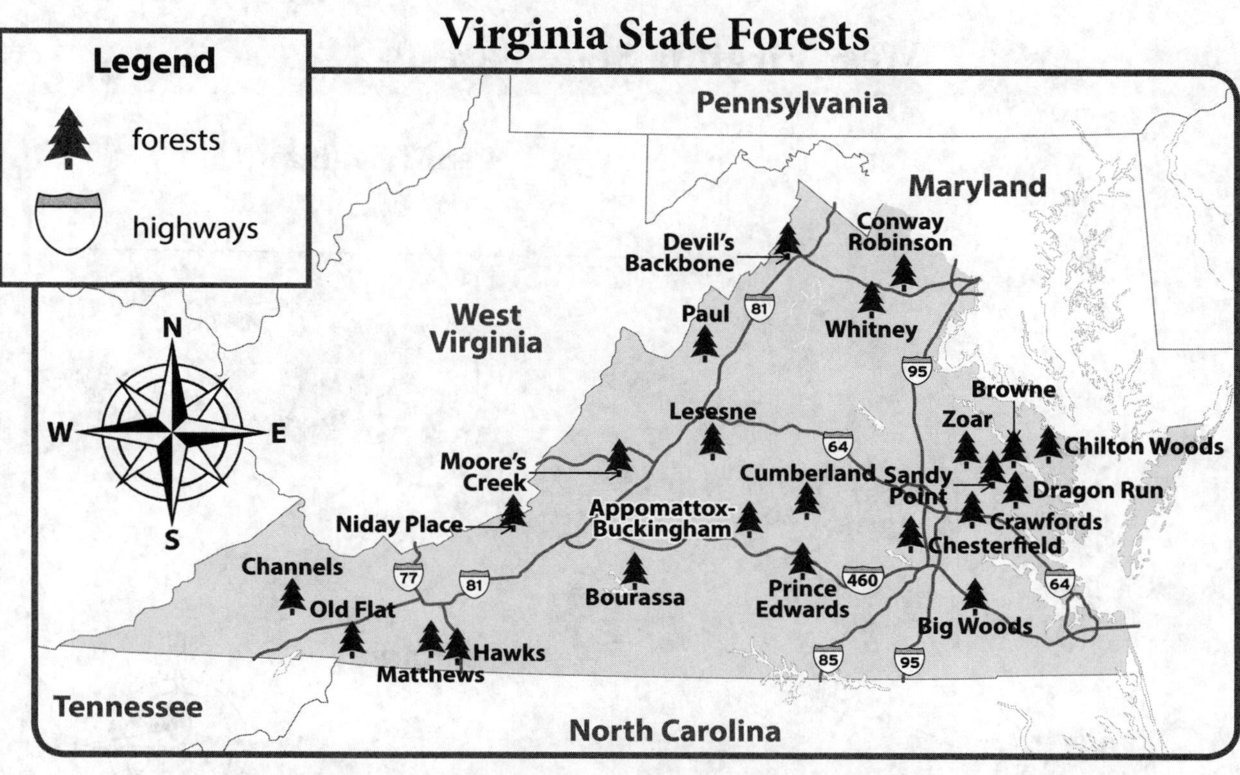

1. Is the Appomattox-Buckingham state forest north or south of Route 460? How do you know?

____

____

2. Which three state forests are the farthest south?

____

____

3. How many state forests are in Virginia?

____

4. Which forest is the farthest west?

____

Creating Maps

Name: ______________________ **Date:** ______________

Directions: Use the clues to label the forests' names on the map.

West Virginia State Forests

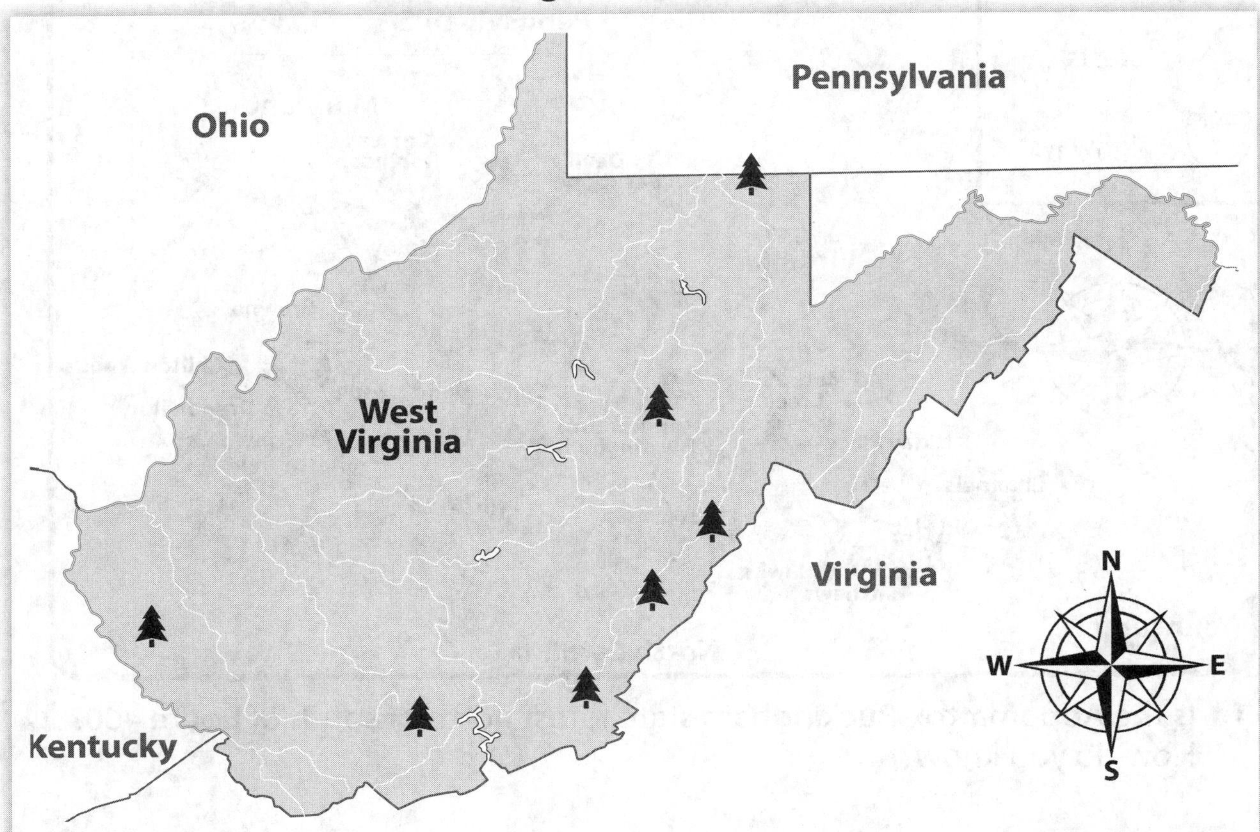

1. Coopers Rock is the farthest north.
2. Cabwaylingo is the farthest west.
3. Kumbrabow is between Coopers Rock and Seneca.
4. Seneca is the farthest east.
5. Camp Creek is the farthest south.
6. Calvin Price is the next forest southwest from Seneca.
7. Greenbrier is northeast from Camp Creek.

Name: ____________________ Date: ____________

Directions: Read the text, and study the map. Then, answer the questions.

Coopers Rock State Forest

Virginia and West Virginia are home to many state forests. The largest in West Virginia is Coopers Rock State Forest. It is in the northern part of the state.

Coopers Rock is used in many ways. The West Virginia Division of Forestry is a group that directs this. This group's goal is to preserve, or maintain, the forest. They plant new trees. They make sure Coopers Lake is stocked with plenty of fish. Scientists research the plants and animals. Their goal is to ensure the wildlife's survival.

Many people visit Coopers Rock each year. Visitors use trails for hiking and biking. Some people camp there overnight. They can fish and hunt at certain times, too. These are just a few of the many benefits of this state forest.

1. What are two things visitors can do at Coopers Rock?

____

____

2. How does the West Virginia Division of Forestry maintain the forest?

____

3. What does *stocked* mean as it is used in the text?

____

____

____

____

Think About It

Name: ______________________ **Date:** ______________

Directions: Scientists sometimes study the ages of trees. They can tell the age of a tree based on the number of rings. Each ring shows one year of growth. Study the diagram, and answer the questions.

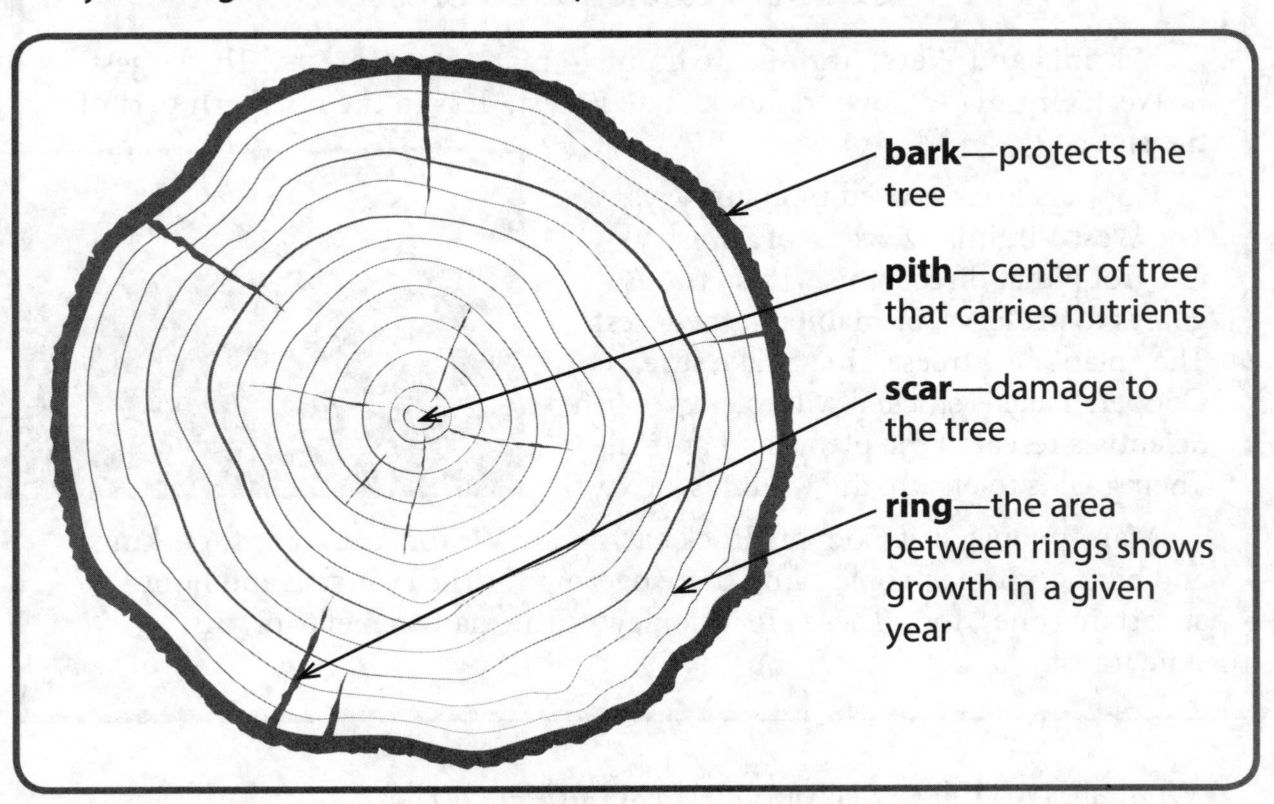

1. How old is this tree? How do you know?

____

____

2. What do you think may have caused the scars on this tree?

____

____

3. Between which two years did the least growth occur? Why do you think there was less growth during this time?

____

____

Name: ______________________________ **Date:** ____________________

Directions: What kinds of trees are found near your home? Draw the different types of trees. Add labels if you know the names of the types of trees.

Reading Maps

Name: ______________________ Date: ______________

Directions: Each state is divided into sections called *counties*. Study the map showing some of Georgia's counties. Then, answer the questions.

Georgia Counties

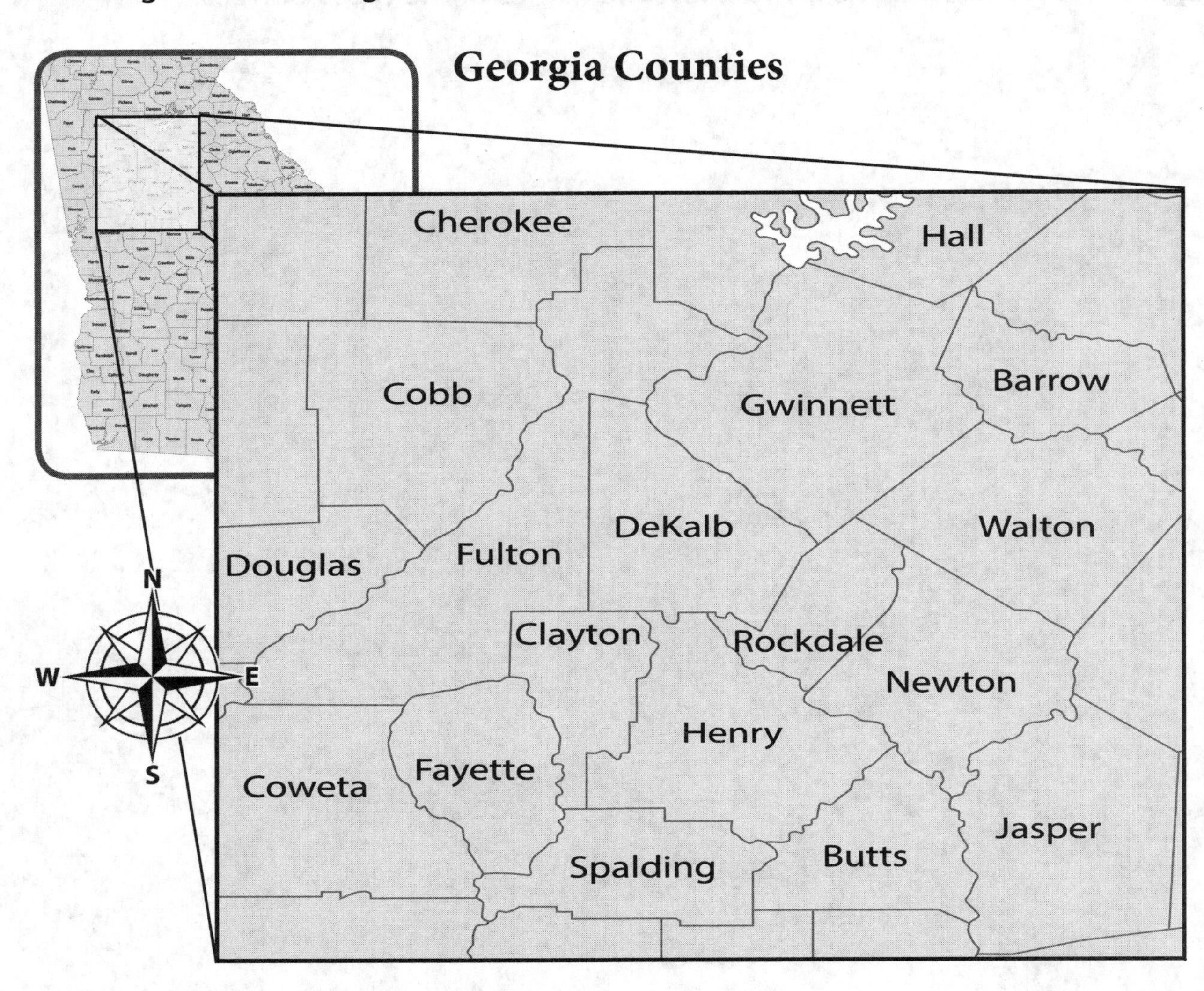

1. Which county shown is the largest?

____

2. How many counties border the county from question 1?

____

3. Which direction would you have to travel to go from Douglas County to Gwinnett County?

____

Name: ______________________ **Date:** ______________

Directions: There are many cities in each of Georgia's counties. Find each city on the map. Use the chart to label the county of each city on the map.

Creating Maps

Georgia Counties

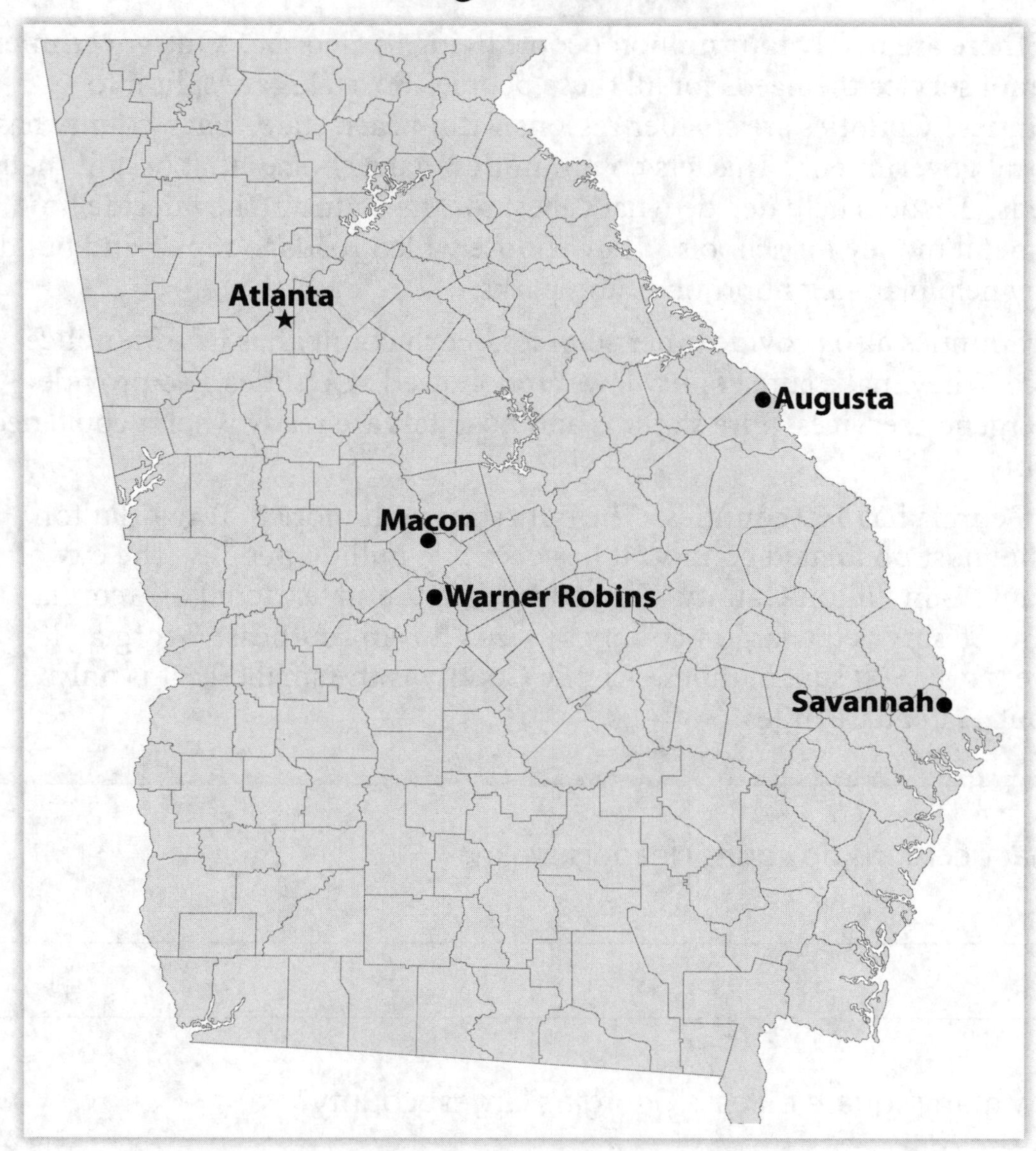

Cities	Counties
Atlanta	Fulton
Augusta	Richmond
Savannah	Chatham
Macon	Bibb
Warner Robins	Houston

Name: ______________________ **Date:** ____________

Read About It

Directions: Read the text. Then, answer the questions.

Georgia Counties

There are nearly four million people living in Georgia. One government cannot service the needs for all these people. So, states are split into counties. Counties are smaller regions within each state. Each county has a local government. This lets communities make choices that best fit their needs. Leaders help decide what is best for the county. They decide how to spend money for schools. They choose which roads to repair and build. They help plan neighborhoods and parks.

Counties also provide other services. Each county has its own police force. They make sure county laws are followed. Counties also provide emergency services. Fire stations and hospitals are ready when people need them.

Georgia has 159 counties. The only state with more is Texas. Fulton is the most populated county. It has over one million people. The city of Atlanta is in Fulton County. This city is also the state capital of Georgia. Ware County is the largest county by size. It is in southeast Georgia. It covers over 900 square miles. Clarke County is the smallest. It is only about 121 square miles.

1. What decisions do county leaders make?

____

____

2. How many square miles is Georgia's largest county?

____

3. Why do you think Fulton County is the most populated?

____

____

Name: ______________________ Date: ____________

Directions: Use the chart to answer the questions.

Georgia's 10 Most Populated Counties in 2016		
Rank	**County**	**Population**
1	Fulton County	1,023,336
2	Gwinnett County	907,135
3	Cobb County	748,150
4	DeKalb County	740,321
5	Chatham County	289,082
6	Clayton County	279,462
7	Cherokee County	241,689
8	Henry County	221,768
9	Forsyth County	221,009
10	Richmond County	201,647

1. How many more people live in Cobb County than Richmond County?

____

2. Which two counties are closest in population?

____

3. Why might a county have more people than other counties?

____

____

Name: ______________________ Date: ______________

Directions: Think about the county where you live. List three things you like about it. List three things that can be improved.

Things I Like	Ways to Improve

Name: ______________________ Date: ____________

Directions: One of the longest rivers in the United States is the Mississippi River. It flows through 10 states. Study the map, and answer the questions.

1. Which states border the Mississippi River to the east?

____

____

____

____

____

____

____

2. What major body of water does the Mississippi River flow into?

____

3. Which state has the least amount of the Mississippi River as its border?

____

____

____

____

____

Creating Maps

Name: ______________________________ Date: ______________

Directions: Follow the steps to complete the map.

1. Trace the Mississippi River in blue.

2. Use symbols to label the Arkansas River, the Illinois River, and the Tennessee River. Each of these rivers is located in the state with the same name.

3. Create a legend to show what your symbols mean.

Name: ______________________ Date: ______________

Directions: Read the text, and study the photo. Then, answer the questions.

Rivers in Alabama

There are many rivers in Alabama. People have relied on them for hundreds of years. Explorers and tribes traveled by river. They used it for drinking water and fishing, too.

In the early 1900s, the steamboat was invented. This changed how people used rivers. Steamboats made river travel much faster. Trips that took days now took hours. People were able to settle all over Alabama. They built cities and towns along the rivers.

Crops were shipped by steamboat to other cities. Cotton is one of Alabama's largest resources. Steamboats took this important crop to be sold in other cities and towns. This helped boost the state's economy.

1. How did American Indian tribes use the river?

____

____

2. How did steamboats change river travel?

____

____

3. How did steamboats help the state's economy?

____

____

____

Name: ______________________ Date: ______________

Directions: There are many river ports in Mississippi. A *port* is an area near a town or city where ships can dock, or park. Use the map of the ports in Mississippi to answer each question.

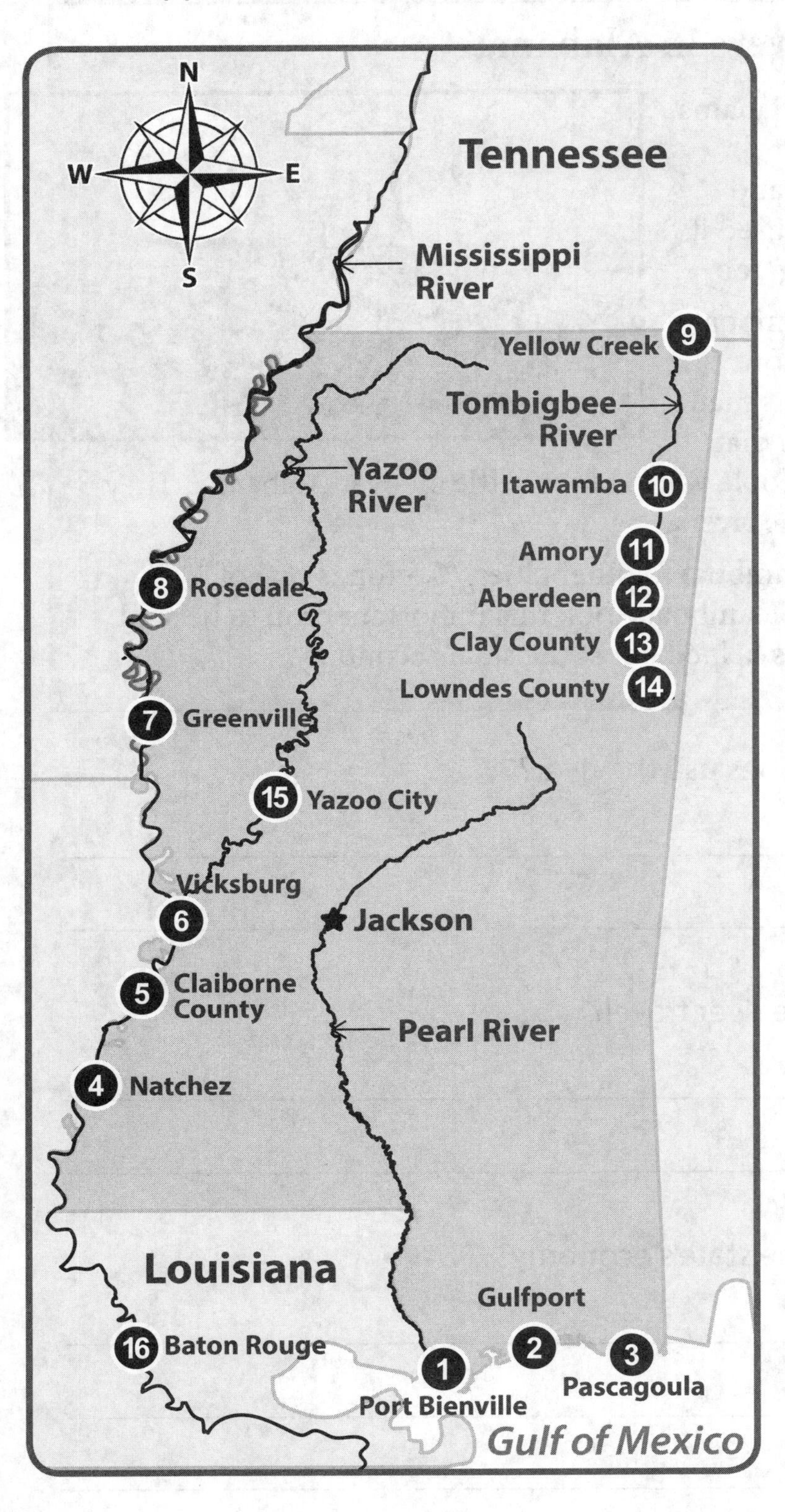

1. What do the locations of ports 4 through 8 have in common?

____

____

____

____

2. Which port would a ship traveling north from Baton Rouge arrive at next?

____

____

____

3. Which port might a ship from another country dock at? Explain your thinking.

____

____

____

____

____

____

____

Name: ______________________ Date: ______________

Directions: Complete the Venn diagram to compare and contrast road travel and river travel.

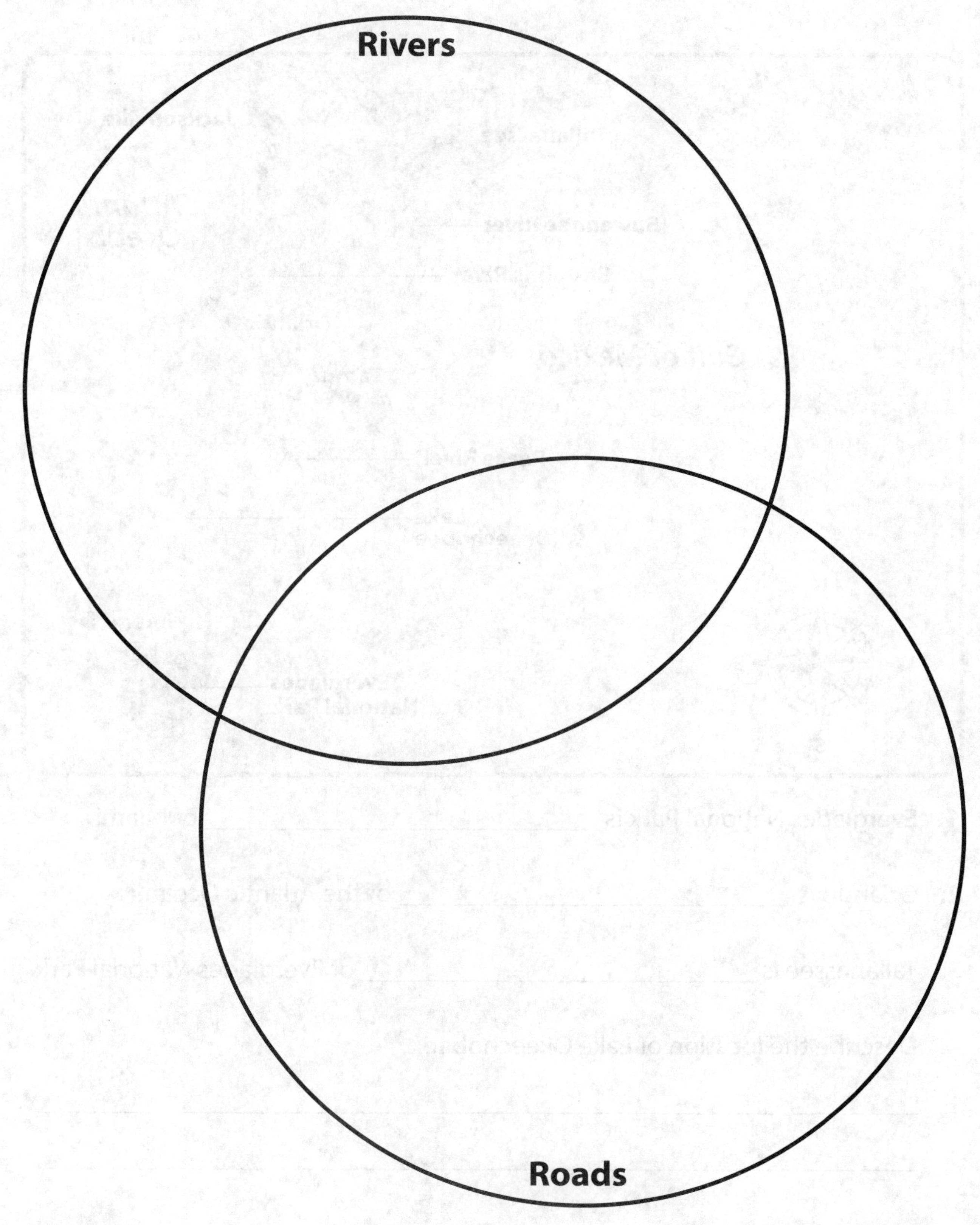

Name: ______________________ Date: ______________

Reading Maps

Directions: Study the map. Then, use the compass rose to answer the questions.

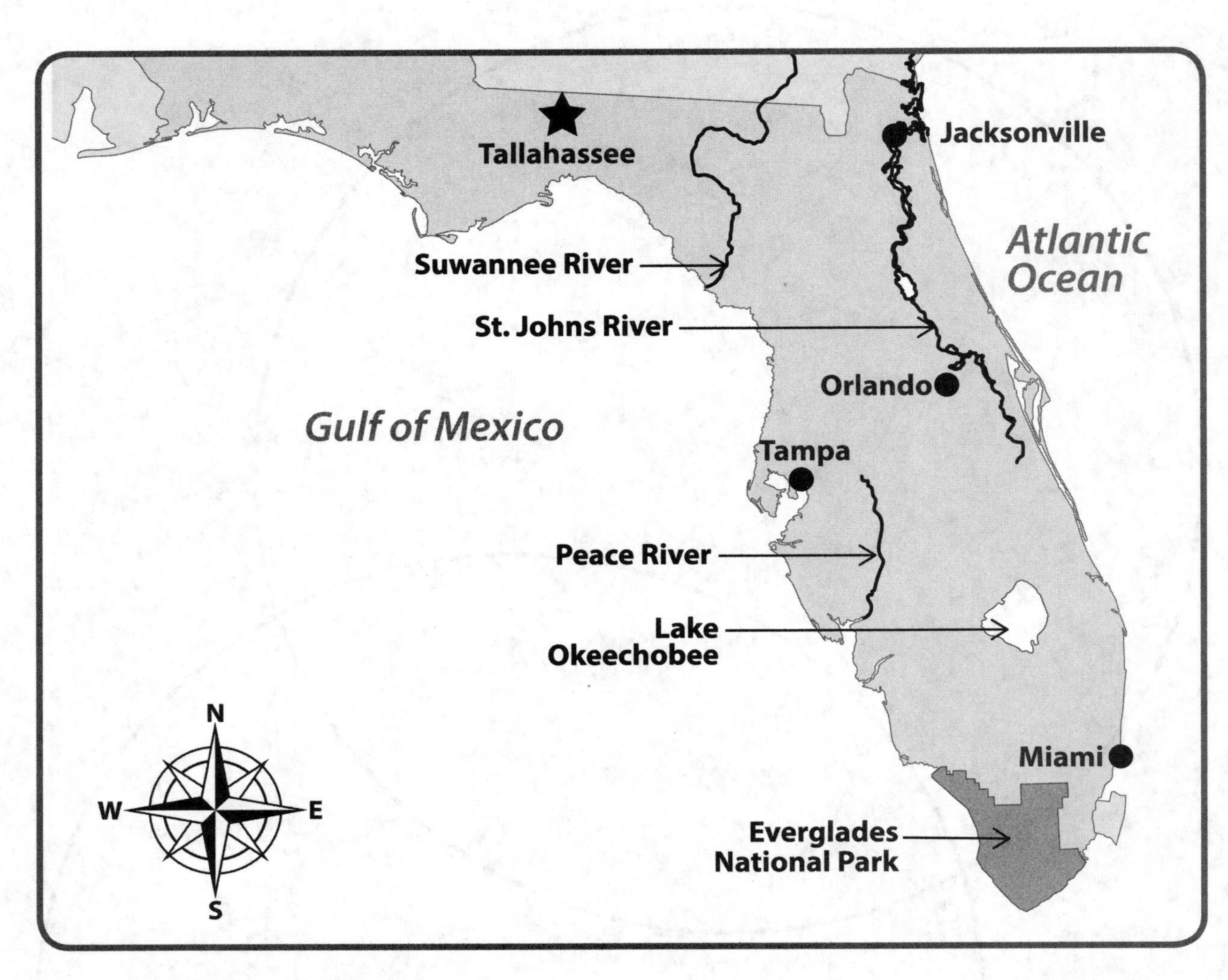

1. Everglades National Park is ______________________ of Miami.

2. Orlando is ______________________ of the Atlantic Ocean.

3. Tallahassee is ______________________ of Everglades National Park.

4. Describe the location of Lake Okeechobee.

____

____

Name: ______________________________ Date: ________________

Directions: Use the clues to label the cities on the map.

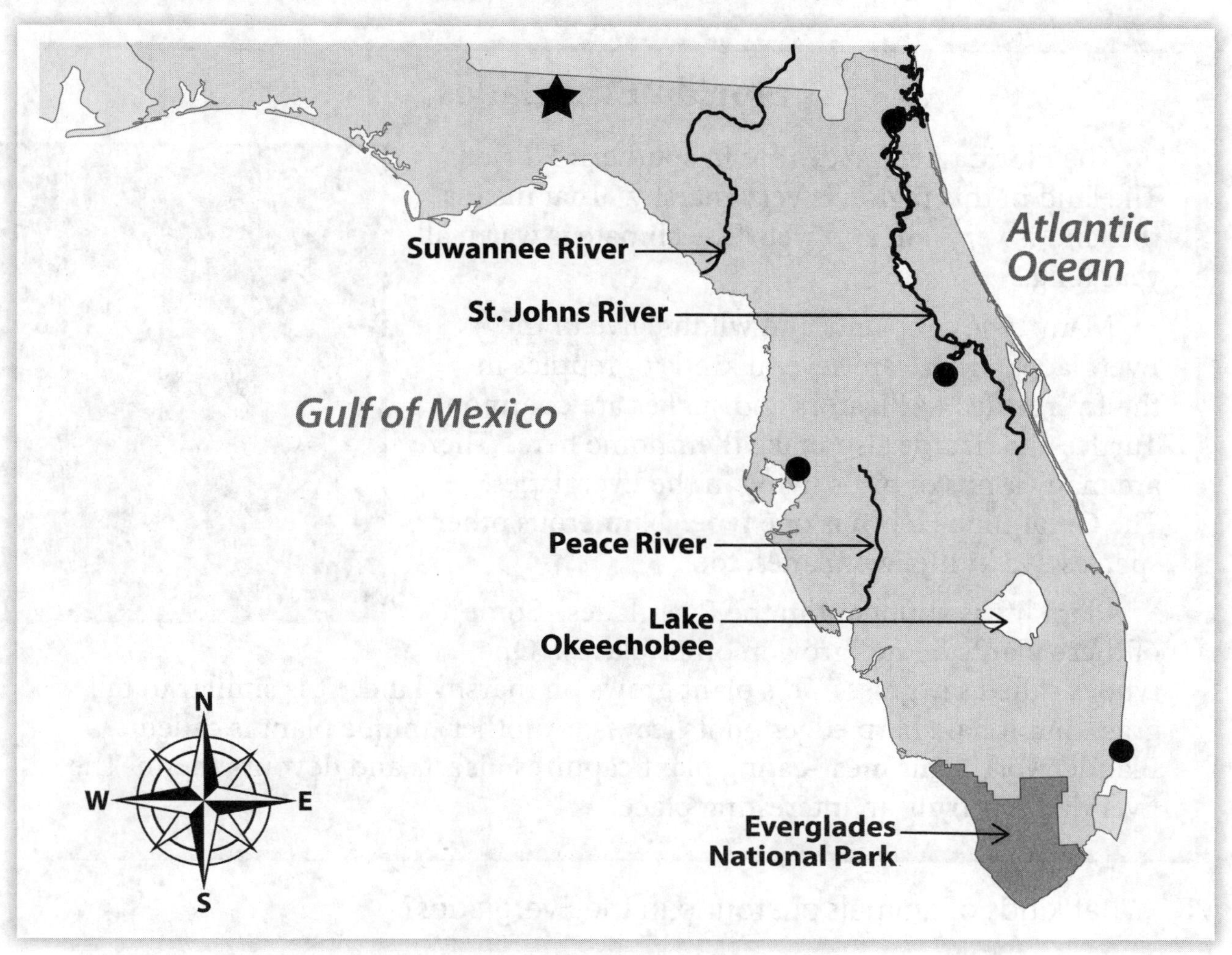

1. Tallahassee is the state capital. It is in the northwestern part of the state.

2. Orlando is west of the St. Johns River.

3. Miami is northeast of Everglades National Park.

4. Jacksonville is east of Tallahassee.

5. Tampa is northwest of the Peace River.

Read About It

Name: ______________________ **Date:** ______________

Directions: Read the text, and study the photo. Then, answer the questions.

Florida Everglades

The Florida Everglades are in southern Florida. The land in this region is very marshy. That means the land is very soft and wet. The climate is warm all year long.

Many types of plants and wildlife live in the Everglades. There are several kinds of reptiles in the Everglades. Alligators and snakes are common. Turtles and lizards also make their home here. There are many kinds of birds living in the Everglades. The Great Blue Heron is one type. Numerous other species live in the Everglades, too.

Plant life is abundant in the Everglades. Some of these plants do not grow in other places. One type is called sawgrass. This plant grows on marshy land. It is similar to tall grass, but it has sharp edges on its leaves. Another unique plant is called Bladderwort. This meat-eating plant captures insects and devours them. The Everglades is truly an interesting place!

1. What kinds of animals are found in the Everglades?

____

____

2. Why do you think one plant is called sawgrass?

____

____

3. What is a Bladderwort?

____

____

Name: ______________________ Date: ______________

Directions: People have released Burmese pythons into the Everglades. These snakes disrupt the food chain since they are not from the region. Scientists work hard to remove them. Study the bar graph, and answer the questions.

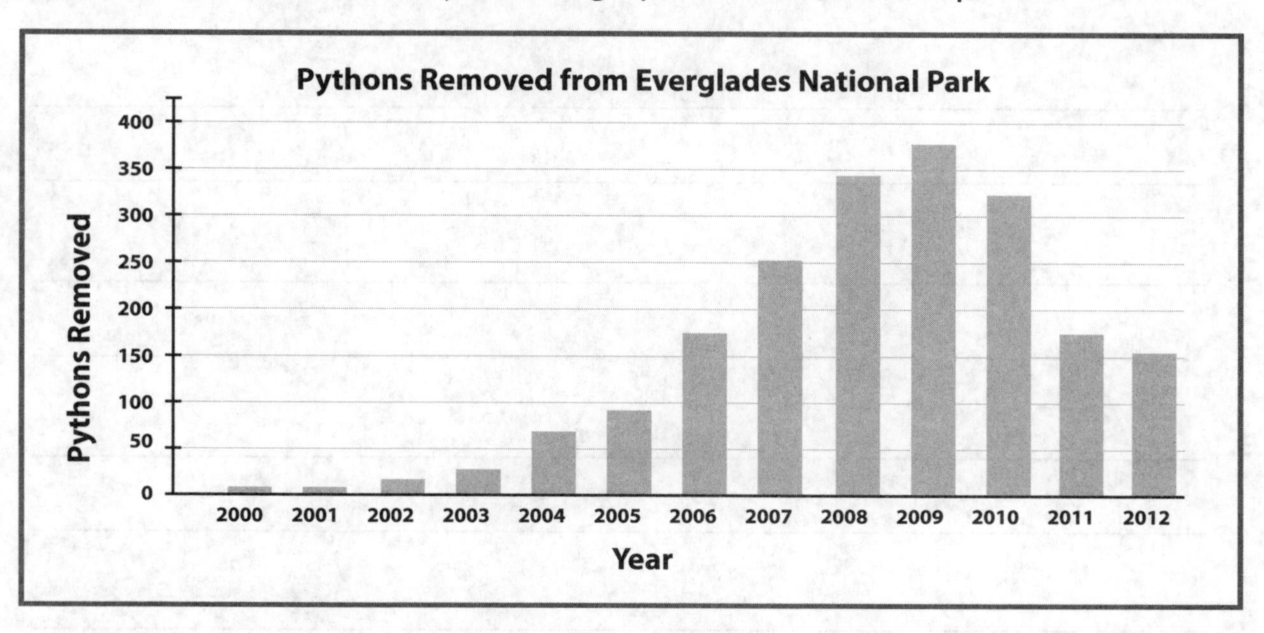

1. About how many pythons were removed between 2010 and 2012?

____

2. Why do you think the number of pythons caught increased between 2000 and 2009?

____

____

____

3. Why do you think the number of caught pythons decreased between 2009 and 2012?

____

____

____

Name: ______________________ **Date:** ______________

Directions: Imagine you are taking a tour of the Florida Everglades. What questions would you ask the tour guide? Write five questions that you would like answered. Then, draw a picture of the Everglades.

1. ______________________

2. ______________________

3. ______________________

4. ______________________

5. ______________________

Name: ______________________________ Date: ______________

Directions: Study the map, and answer the questions.

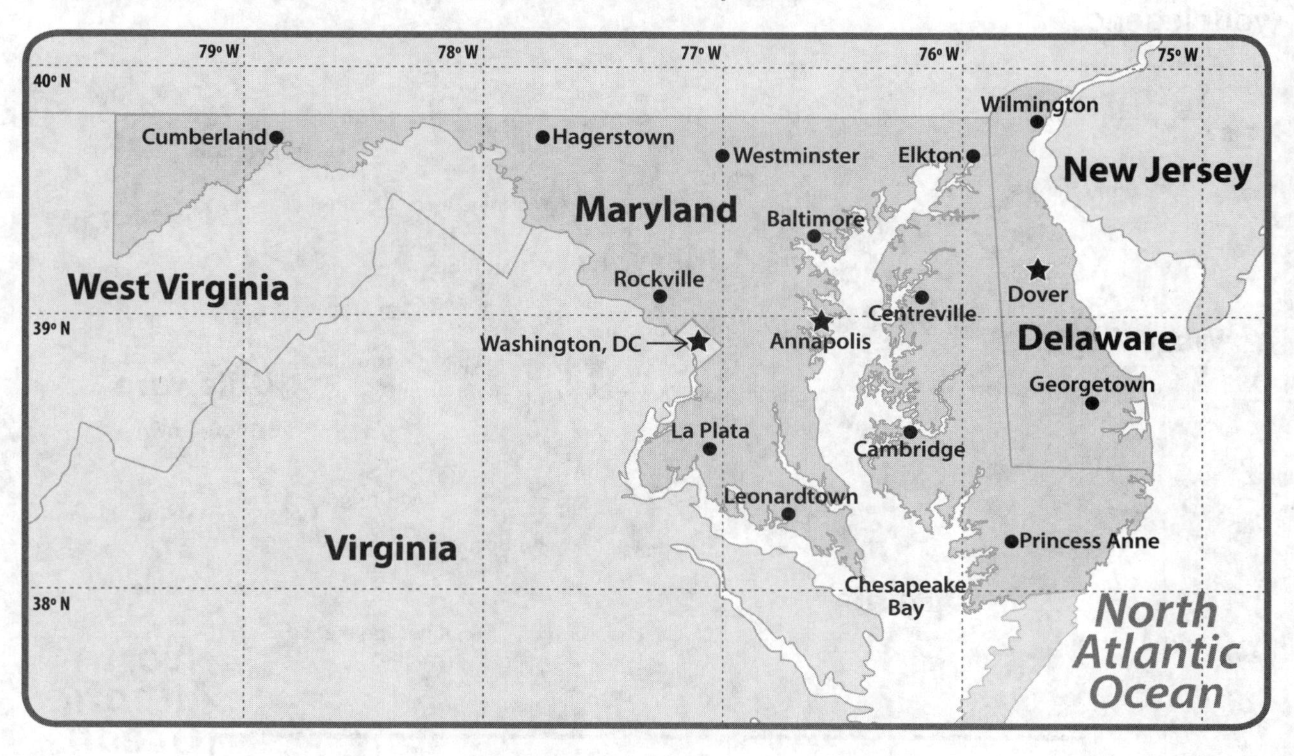

1. What city is located at 39° North latitude?

____

2. Place an *X* on 39°N, 75 °W. What is at these coordinates?

____

3. Name two cities near 77° West longitude.

____

4. Give the approximate coordinates for Elkton, Maryland.

____

5. Give the approximate coordinates for Centreville, Maryland.

____

Name: ______________________ **Date:** ______________

Directions: Create a legend for the map. Add each of the items listed below to your legend.

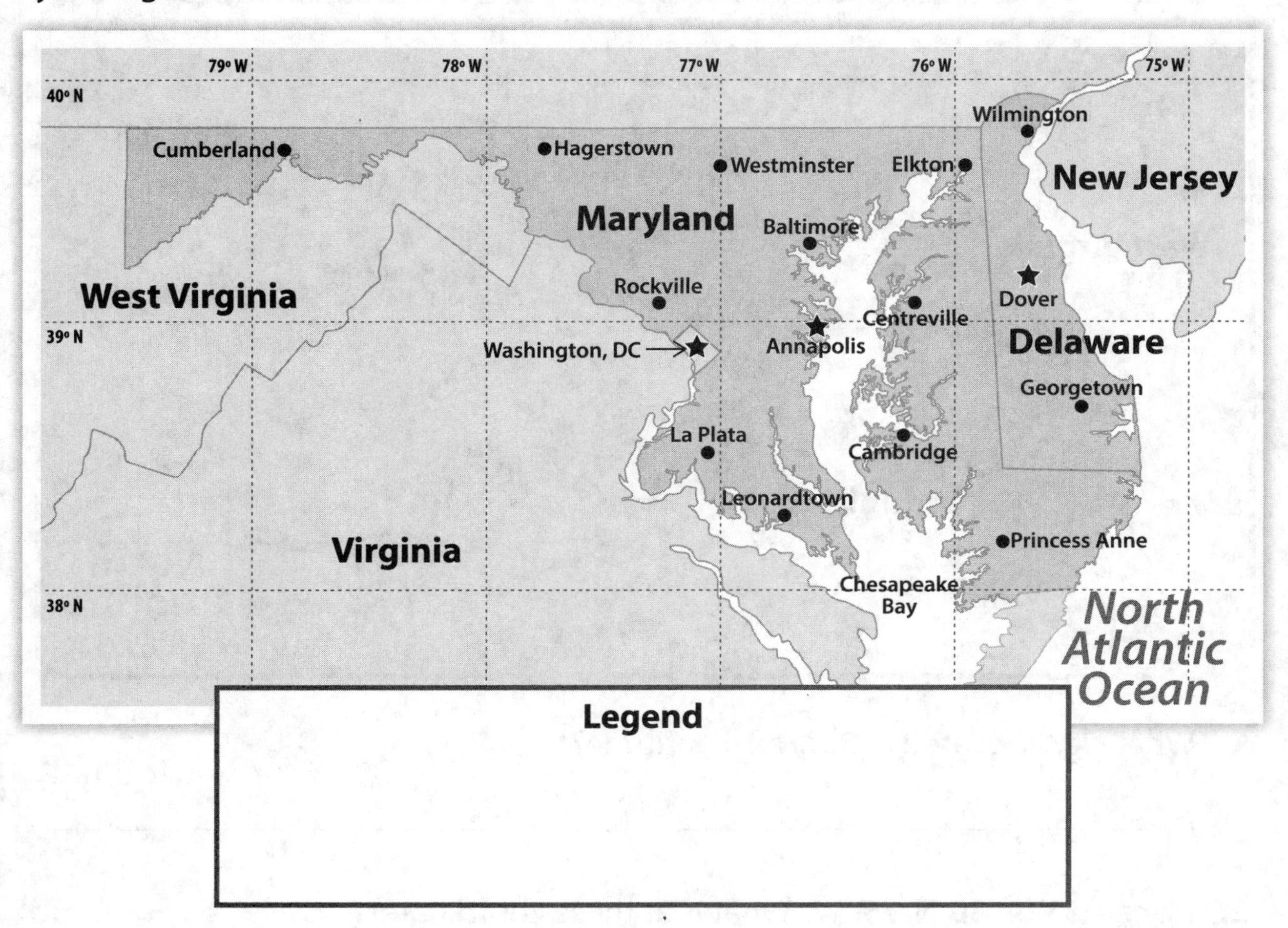

1. Capitals are shown with stars.
2. Major cities are shown with dots.
3. State borders are shown with gray lines.
4. Latitude and longitude lines are shown with thin, dotted lines.

Challenge: Cover the names of the states with sticky notes. Write the state names on the sticky notes. Then, check your answers.

Name: ______________________ Date: ______________

Directions: Read the text, and study the photo. Then, answer the questions.

The Atlantic Ocean

Delaware and Maryland border the Atlantic Ocean. The ocean is a huge resource for these states.

One key ocean resource is food. The ocean is home to many types of fish. There are also shellfish, such as crabs, lobsters, and clams. Delaware and Maryland rely on this seafood. It is a big part of their economies. It provides jobs for many people.

The ocean also brings many tourists to Maryland and Delaware. Millions of people vacation in these states. Ocean City is a popular beach in Maryland. Dewey Beach attracts many people to Delaware.

People use sailboats near the beaches. Large ships transport goods. Cruise ships dock in Baltimore, Maryland. These boats carry thousands of people across the ocean.

Baltimore, Maryland

1. What are two beaches in Maryland and Delaware?

____

2. What are two things boats are used for on the Atlantic Ocean?

____

____

3. Why might people build homes and businesses near the ocean?

____

____

____

Name: ______________________ Date: ____________

Directions: This line graph shows the temperature of the Atlantic Ocean at Dewey Beach in Delaware. Study the graph. Then, answer the questions.

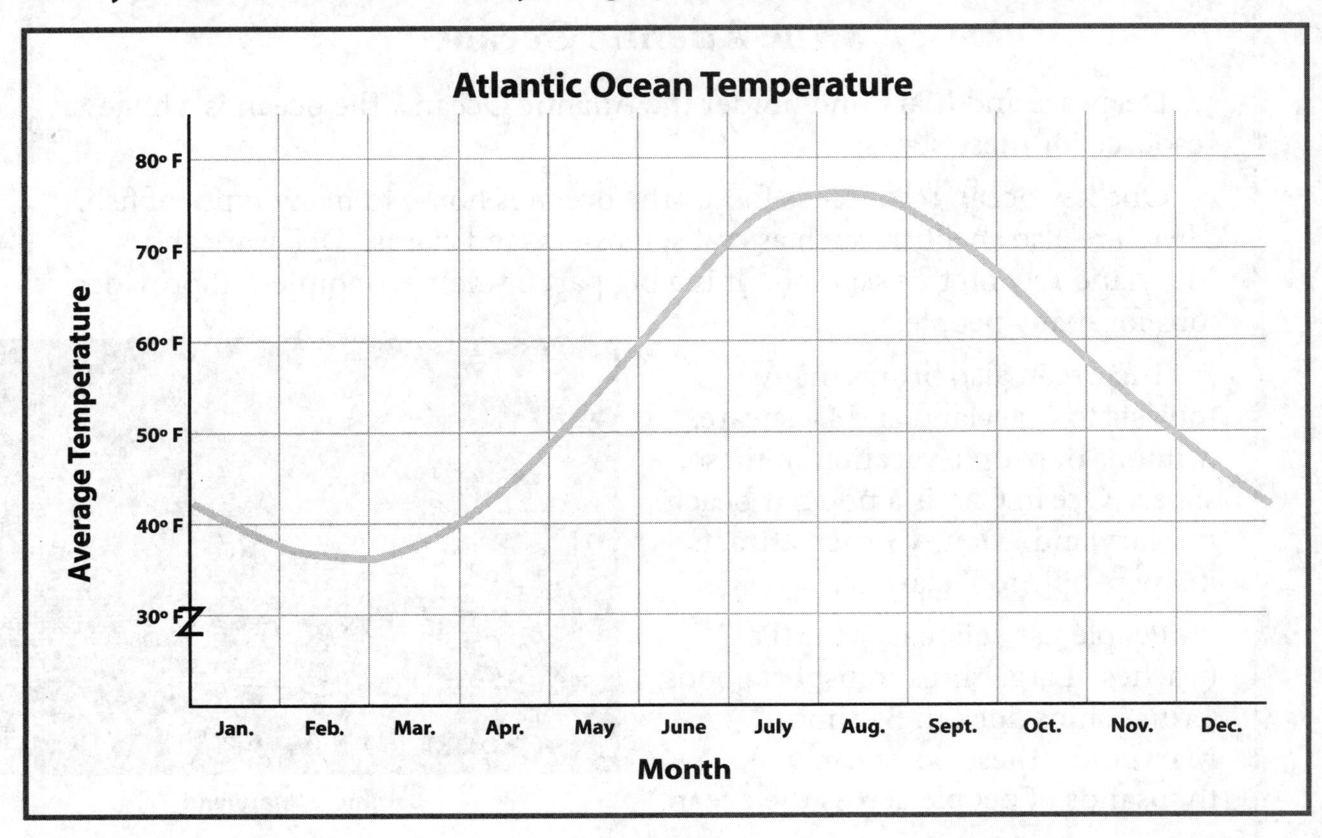

1. During which months is the ocean temperature less than 40 degrees?

____

2. How does the temperature of the water relate to the four seasons?

____

____

3. Water freezes at 32°F. Does the water in the Atlantic Ocean freeze at any point during the year? How do you know?

____

____

Name: ______________________ Date: ______________

Directions: What are some things that you like about the ocean? If you have never been to the ocean, think about what you would like to do and see.

The Ocean

I. Activities to do at the beach

A. ______________________

B. ______________________

C. ______________________

D. ______________________

II. Activities to do in the ocean

A. ______________________

B. ______________________

C. ______________________

D. ______________________

III. Interesting animals in or near the ocean

A. ______________________

B. ______________________

C. ______________________

D. ______________________

Reading Maps

Name: ______________________________ Date: ______________

Directions: This diagram shows how running water can be used to create electricity at a power plant. This is called *hydroelectric energy*. Study the diagram, and answer the questions.

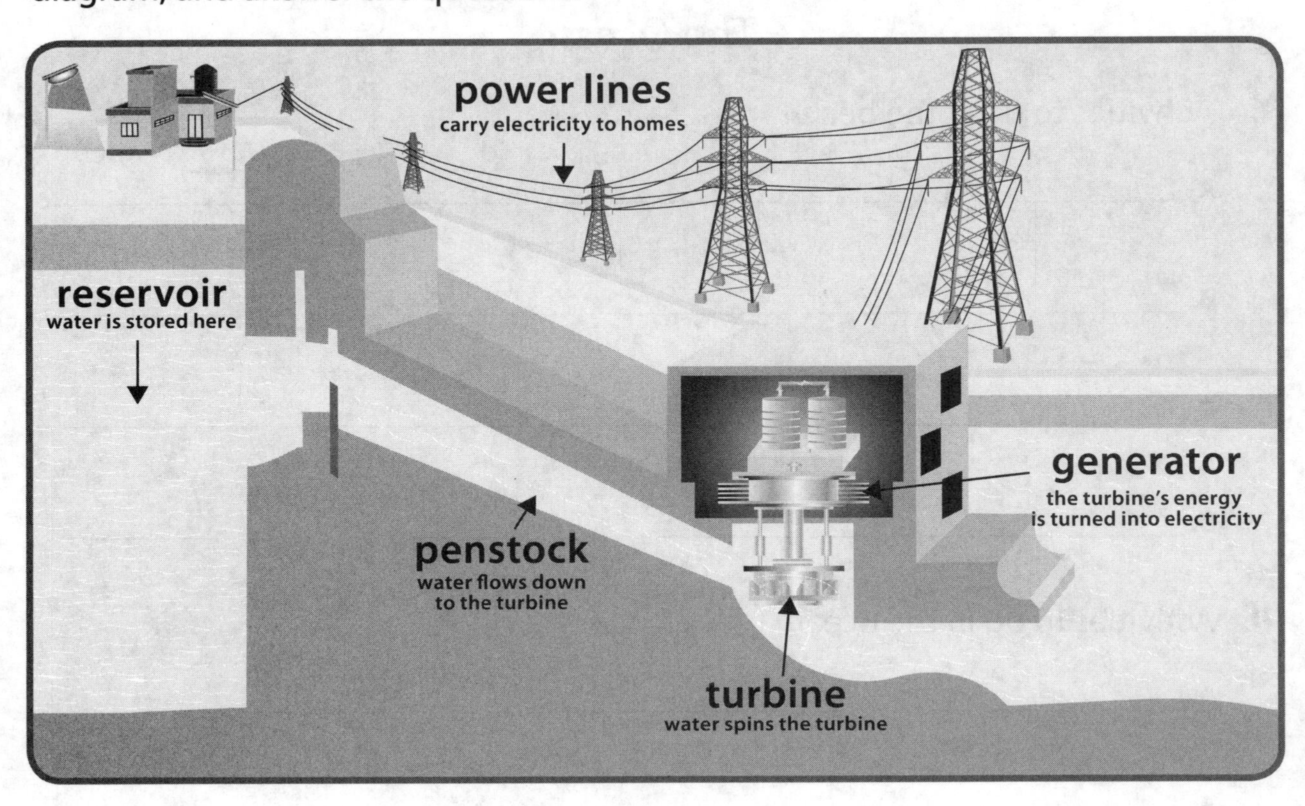

1. Which part stores the water?

____

2. How does water spin the turbine?

____

____

____

3. What is the purpose of the power lines?

____

____

Name: ______________________ Date: ______________

Directions: Renewable energy can be used over and over. Read the definitions of the different types of renewable energy. Use the information to label the diagram.

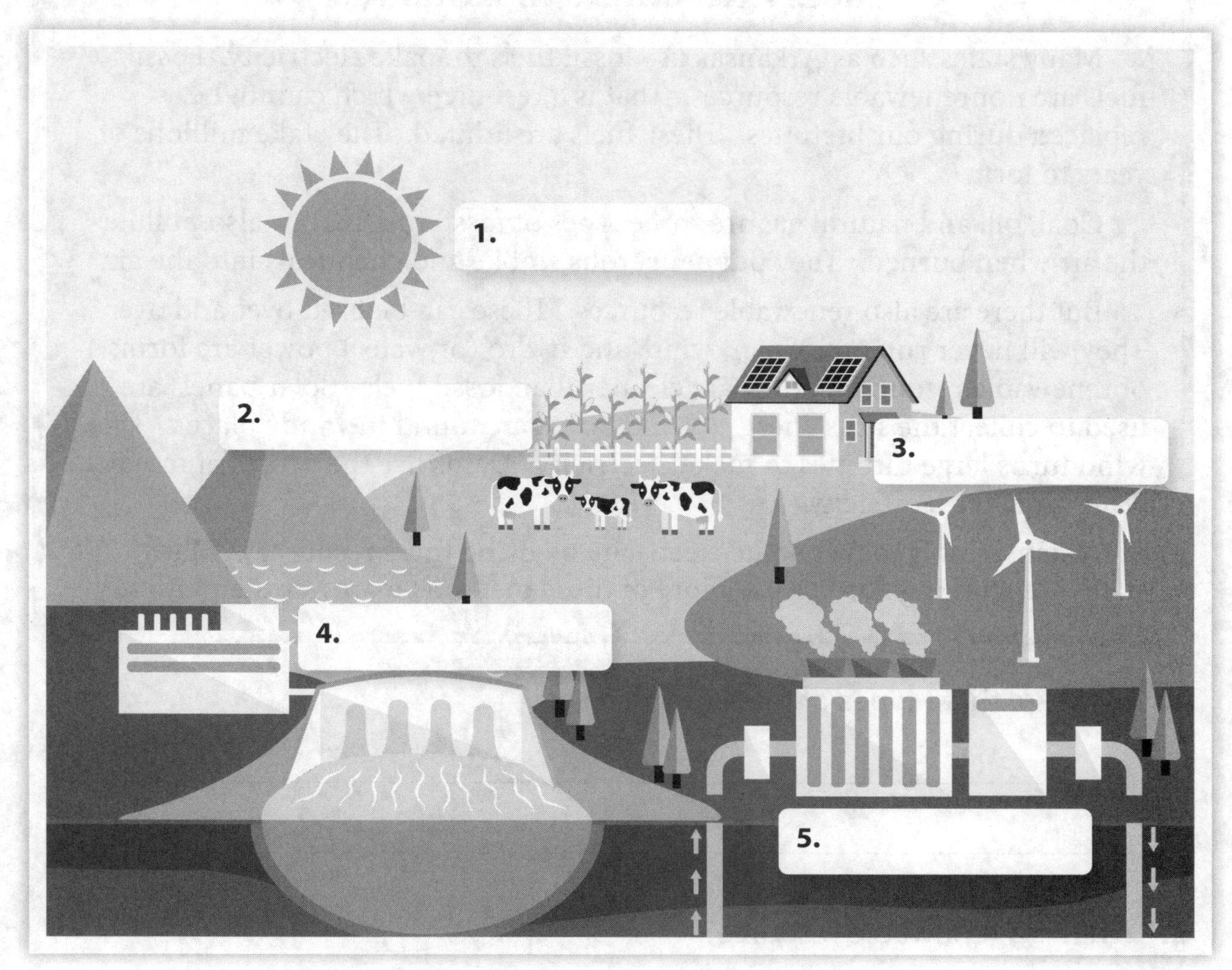

biomass energy—plant or animal matter that can be converted to fuel

geothermal energy—energy from heat inside Earth

hydroelectric energy—energy from moving water

solar energy—energy given off by the sun

wind energy—using moving air to generate electricity

Name: ______________________ Date: ____________

Read About It

Directions: Read the text, and answer the questions.

Energy Resources in Arkansas

Many states such as Arkansas use fossil fuels to make electricity. Fossil fuels are nonrenewable resources. That is a resource which cannot be replaced during our lifetimes. These fuels are limited. They take millions of years to form.

Coal, oil, and natural gas are three types of fossil fuels. They also pollute the air when burned. They put dangerous smoke and chemicals into the air.

But there are also renewable resources. These can be used over and over. They will never run out. Solar, wind, and hydro (or water) power are forms of renewable resources. They are cleaner than fossil fuels. Solar panels are used to collect the sun's energy. Wind farms are found in windy places. The wind turns large turbines to make electricity. Dams use the power of moving water to make electricity.

Arkansas now makes some electricity by using these resources. Many states are working hard to use more of these methods to power their cities.

1. What are three types of fossil fuels?

____

____

2. What is a renewable resource?

____

____

3. How might a coal power plant be harmful to Earth?

____

____

____

Name: ______________________ **Date:** ______________

Directions: This chart shows the amount of natural gas used by 10 states in 2015. Use the chart to answer the questions.

Natural Gas Usage	
state	**cubic feet**
Arkansas	676 billion
Rhode Island	94 billion
Tennessee	313 billion
Florida	1,334 billion
Ohio	969 billion
Texas	4,127 billion
Utah	231 billion
Alaska	337 billion
North Carolina	499 billion
Massachusetts	445 billion

1. Which three states use the most natural gas?

____

2. How many cubic feet do those three states use combined?

____

3. How many states on this chart use less natural gas than Arkansas?

____

4. Why might Texas use more natural gas than any other state on this chart?

____

____

Name: ______________________ Date: ______________

Directions: Write a letter to a local electric company. Convince them to make electricity using more renewable resources.

Dear Electric Company,

____

____

____

____

____

____

____

____

____

____

____

____

____

Sincerely,

____

____

Name: ______________________________ **Date:** ______________

Directions: Many cities in North Carolina and South Carolina are near the coast. These cities can be affected by hurricanes. Use the scale on this map to answer the questions.

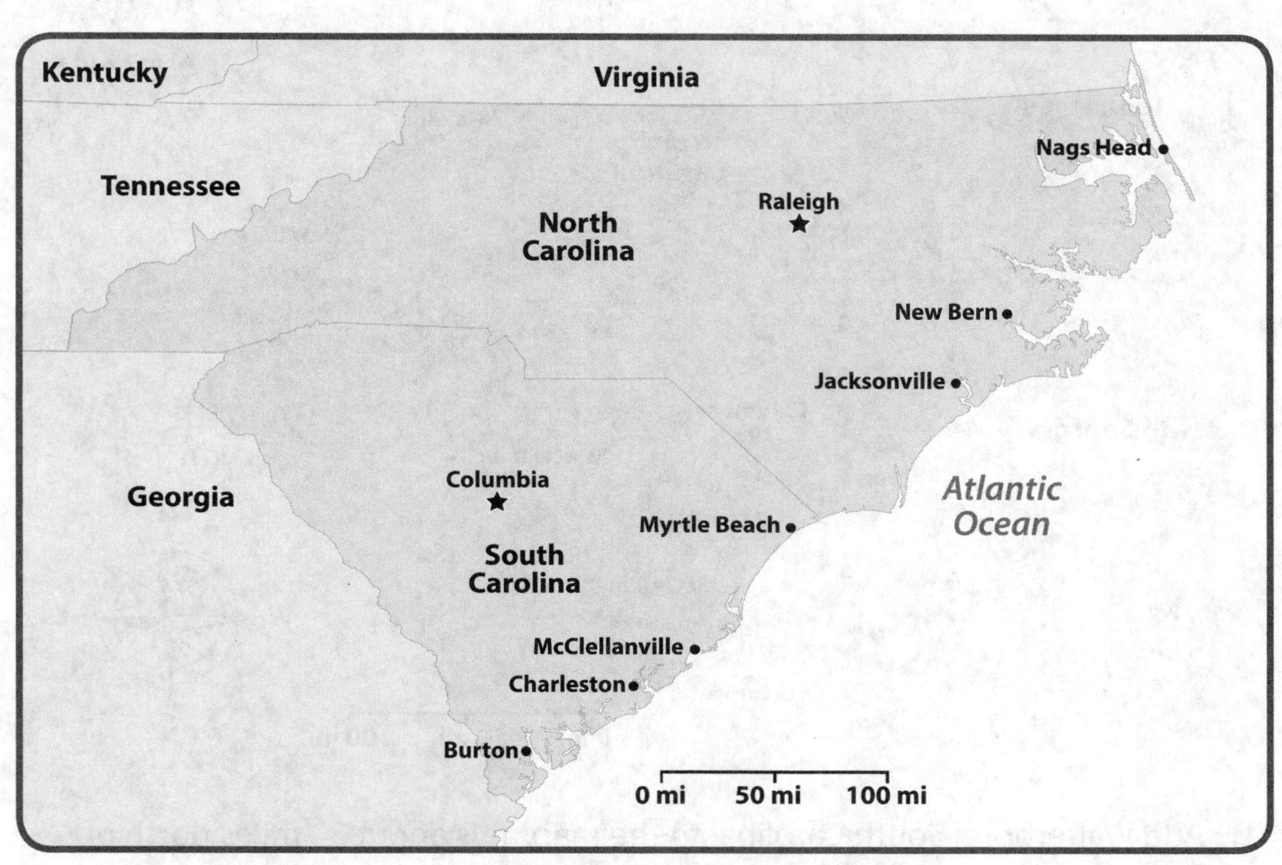

1. About how far is it from Charleston to Myrtle Beach?

____

2. About how many miles is it from Jacksonville to New Bern?

____

3. Is 500 miles a good estimate for the length of the South Carolina coast? Why or why not?

____

____

____

Name: ______________________________ **Date:** ____________________

Directions: Use the scale to add three more cities on the map.

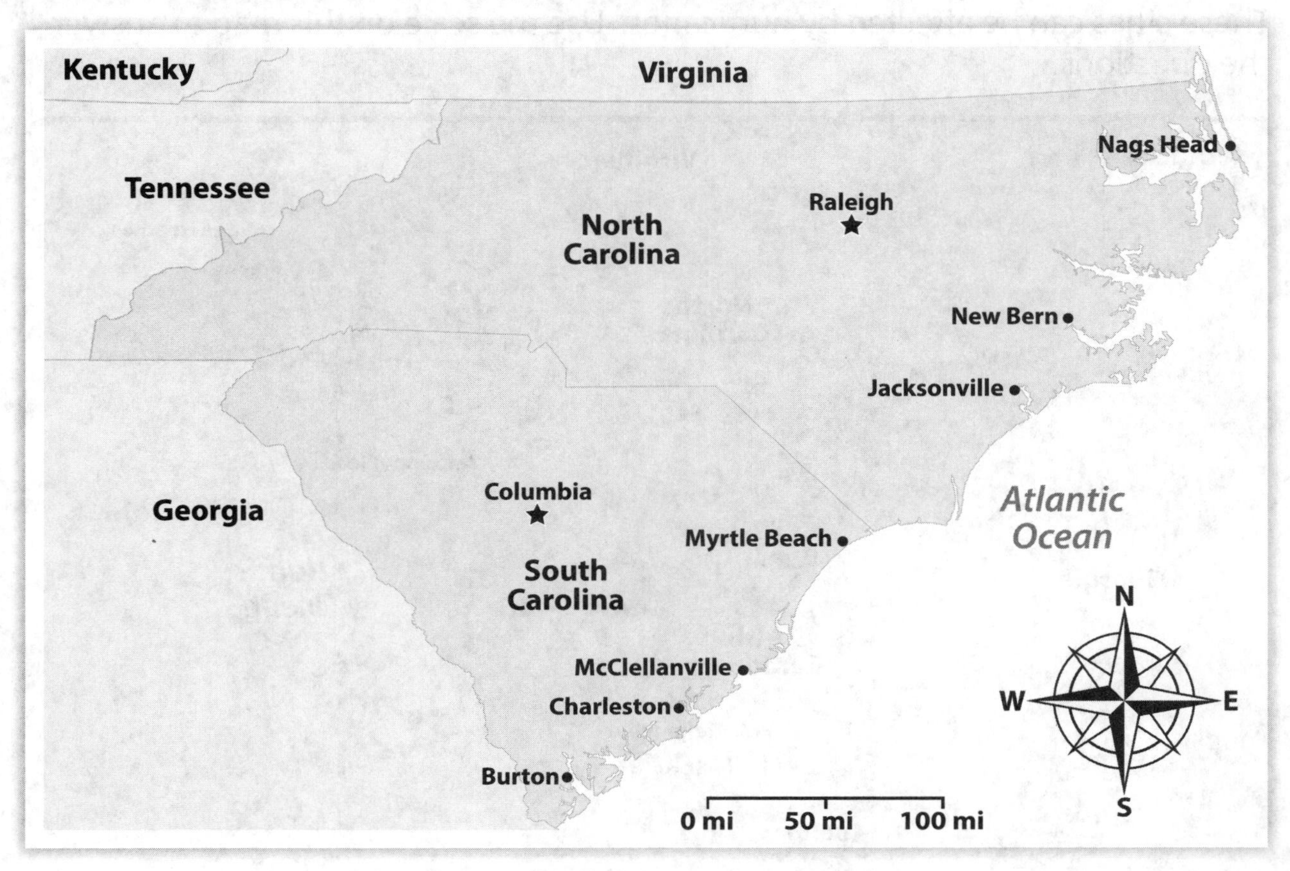

1. Add Walterboro, South Carolina, to the map. It is about 30 miles north of Burton.

2. Add Manning, South Carolina, to the map. It is about 100 miles west of Myrtle Beach.

3. Add Greenville, North Carolina, to the map. It is about 75 miles north of Jacksonville.

Challenge: Look up the actual locations of these cities. Ask an adult to help you. How close were your points on the map above?

Name: ______________________ **Date:** ____________

Directions: Read the text, and study the photo. Then, answer the questions.

Hurricanes

Hurricanes are powerful storms. They contain strong winds and rain. Many hurricanes form in the Atlantic Ocean. The warm waters provide the energy for these storms. As a hurricane forms, it spins counterclockwise. The center of a hurricane is known as the *eye*.

As these storms move, the winds get stronger. Wind speeds can reach over 150 miles (241 kilometers) per hour! Usually, hurricanes move northwest. This puts coastal states, such as North Carolina and South Carolina, at risk. These fierce storms can damage cities in their paths.

The National Weather Service tracks these storms ahead of time. This gives states time to prepare. Every storm is rated on a scale of 1 to 5 based on its wind speeds. A rating of 5 is the strongest. Each storm is also given a name. One of the strongest storms to reach the Carolinas was Hurricane Matthew in 2016. It was a category 5 storm.

1. What gives a hurricane its energy as it forms over the ocean?

____

2. Why are North Carolina and South Carolina at risk for hurricanes?

____

____

3. Circle the eye of the storm in the photo.

4. What can you tell about the size of the hurricane in the photo?

____

Think About It

Name: ______________________ Date: ______________

Directions: Hurricanes are rated on a scale of 1 to 5. This is determined by the speed of the wind. Study the table, and answer the questions.

Rating	Wind Speeds	Damage	Homes	Trees
Category 1	74–95 mph	very dangerous winds that will cause damage	damage to roofs	large branches will snap
Category 2	96–110 mph	extremely dangerous winds with major damage	roof damage with possible broken windows	small trees uprooted
Category 3	111–129 mph	devastating damage will occur	some homes can be destroyed	many trees will be uprooted and block roads
Category 4	130–156 mph	catastrophic damage	many homes may be destroyed	most trees will be uprooted
Category 5	157 mph and up	catastrophic damage	homes destroyed and large buildings severely damaged	nearly all trees will be snapped and uprooted

1. Describe what can happen to homes and trees during a Category 1 hurricane.

____

____

____

2. Why do you think the categories stop at Category 5?

____

____

____

Name: ______________________ Date: ______________

Directions: It is important to be ready when a hurricane is coming. This is a list of recommended emergency items. Select three items you think are most important, and explain why you chose them.

Basic Disaster Supplies Kit

- ✓ **water**
- ✓ **food**
- ✓ **radio**
- ✓ **flashlight**
- ✓ **first aid kit**
- ✓ **whistle to signal for help**
- ✓ **moist towelettes**
- ✓ **tools**
- ✓ **can opener**
- ✓ **local maps**

Reading Maps

Name: ______________________ **Date:** ______________

Directions: Louisiana is one of the wettest states. This map shows the average yearly precipitation for eight states, including Louisiana. Study the map, and answer the questions.

Average Yearly Precipitation

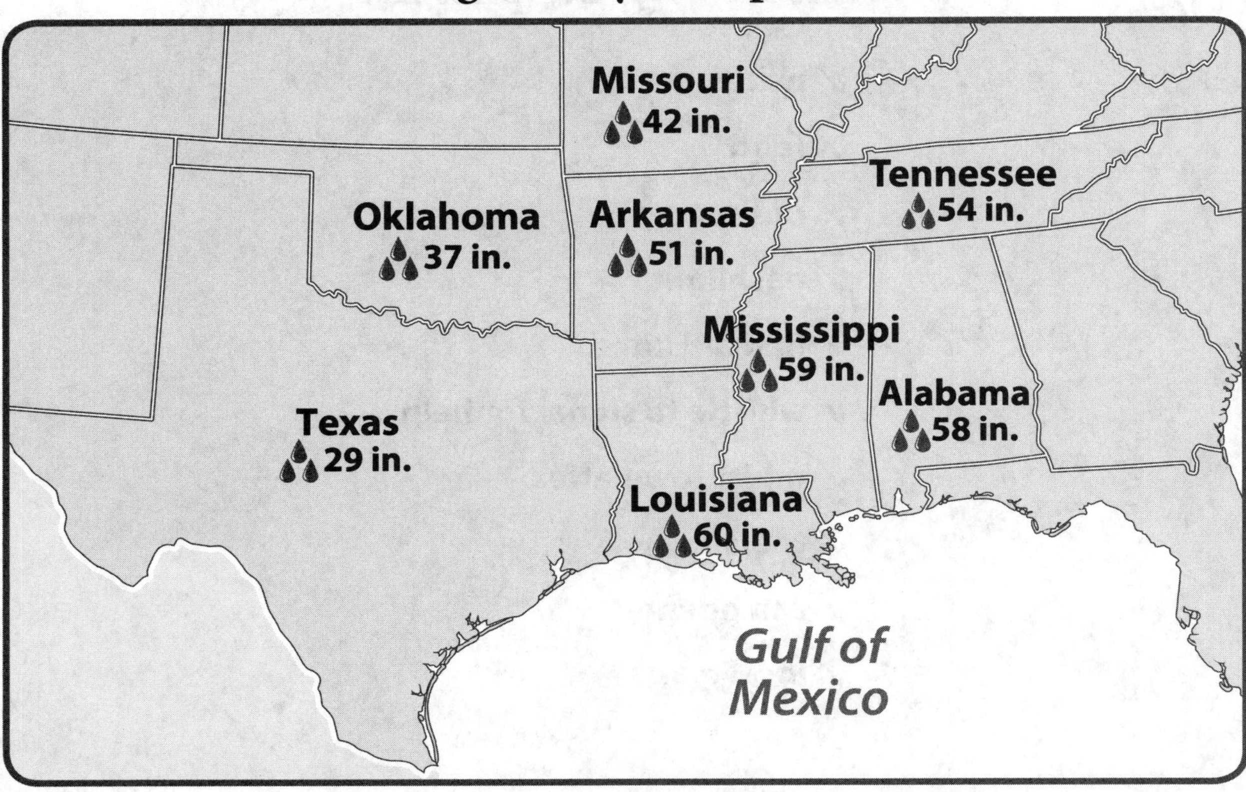

1. About how much precipitation does Louisiana receive per year?

____

2. Which two states receive the least amount of precipitation?

____

3. How much rainfall do Louisiana and Mississippi receive combined?

____

4. How many more inches of precipitation does Louisiana receive than Oklahoma?

____

Name: ______________________________ Date: ________________

Directions: Follow the steps in the box to color the map.

Average Yearly Precipitation

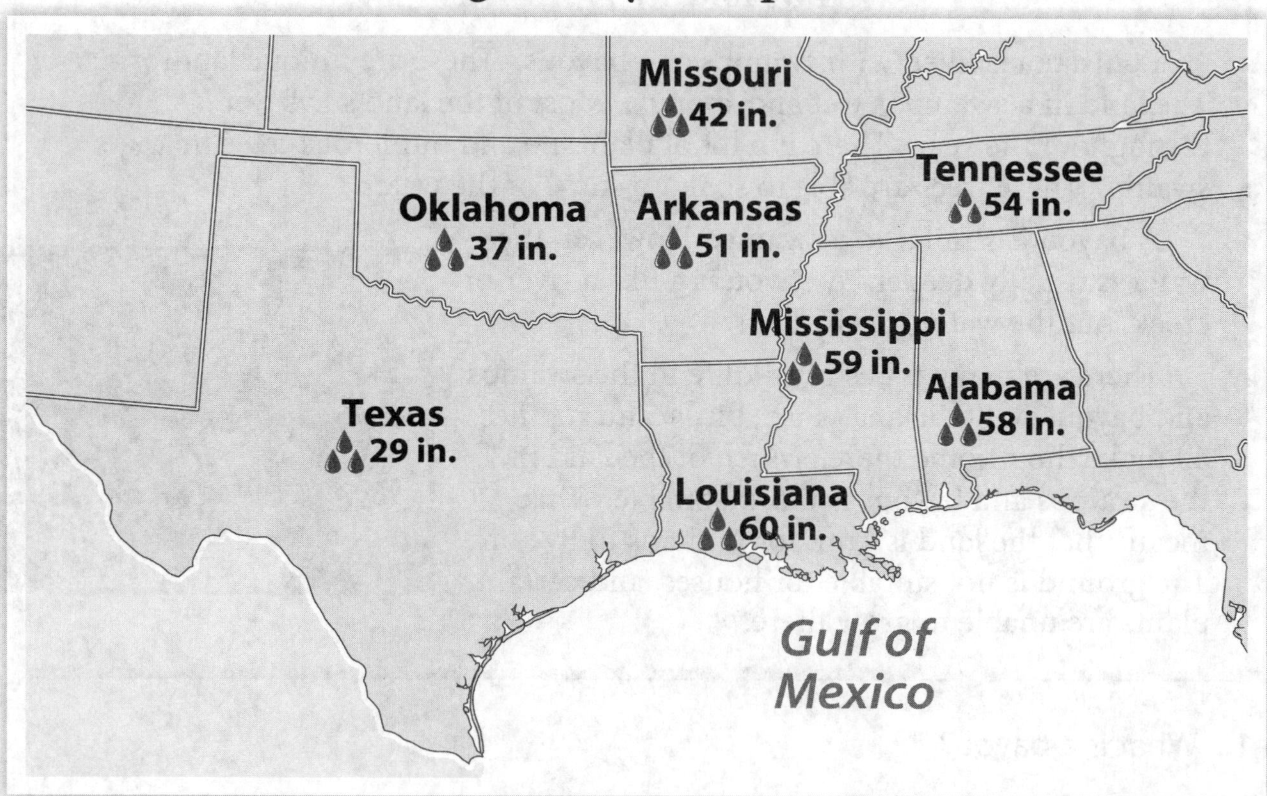

Color states with 20–29 inches yellow.

Color states with 30–39 inches orange.

Color states with 40–49 inches green.

Color states with 50–59 inches blue.

Color states with 60–69 inches purple.

Challenge: Cover the names of the states with sticky notes. Write the state names on the sticky notes. Then, check your answers.

Read About It

Name: ______________________ **Date:** ______________

Directions: Read the text, and study the photo. Then, answer the questions.

Louisiana Wetlands

Louisiana is filled with swamps and bayous. These are unique landforms. The land in a swamp is wet and spongy. Most of the land stays wet throughout the year. There is a lot of plant life and numerous trees in a swamp. These trees are able to soak up some of the water.

A bayou is similar to a swamp. However, the water is usually deeper. A bayou is a like a river or creek, but the water barely flows.

There are many types of wildlife in the swamps and bayous of Louisiana. Fish, birds, and reptiles all make their home there. Much of the land in the swamps and bayous is uninhabitable. That means that the land is unfit for humans to live on. The ground is not suitable for houses, and certain plants are unable to grow there.

1. What is a bayou?

____

____

2. What does *uninhabitable* mean?

____

3. What dangers do you think people might face when traveling through a swamp or a bayou?

____

____

____

____

Name: ______________________ Date: ______________

Directions: This photo shows a house built in a swamp in Louisiana. Study the photo, and answer the questions.

1. How is this house able to stay dry in the swampy waters?

____

____

2. What challenges do you think people living here would face?

____

____

3. What methods of travel might someone living here use?

____

____

Name: ______________________ Date: ______________

Directions: Complete the Venn diagram to compare and contrast your community with a swamp.

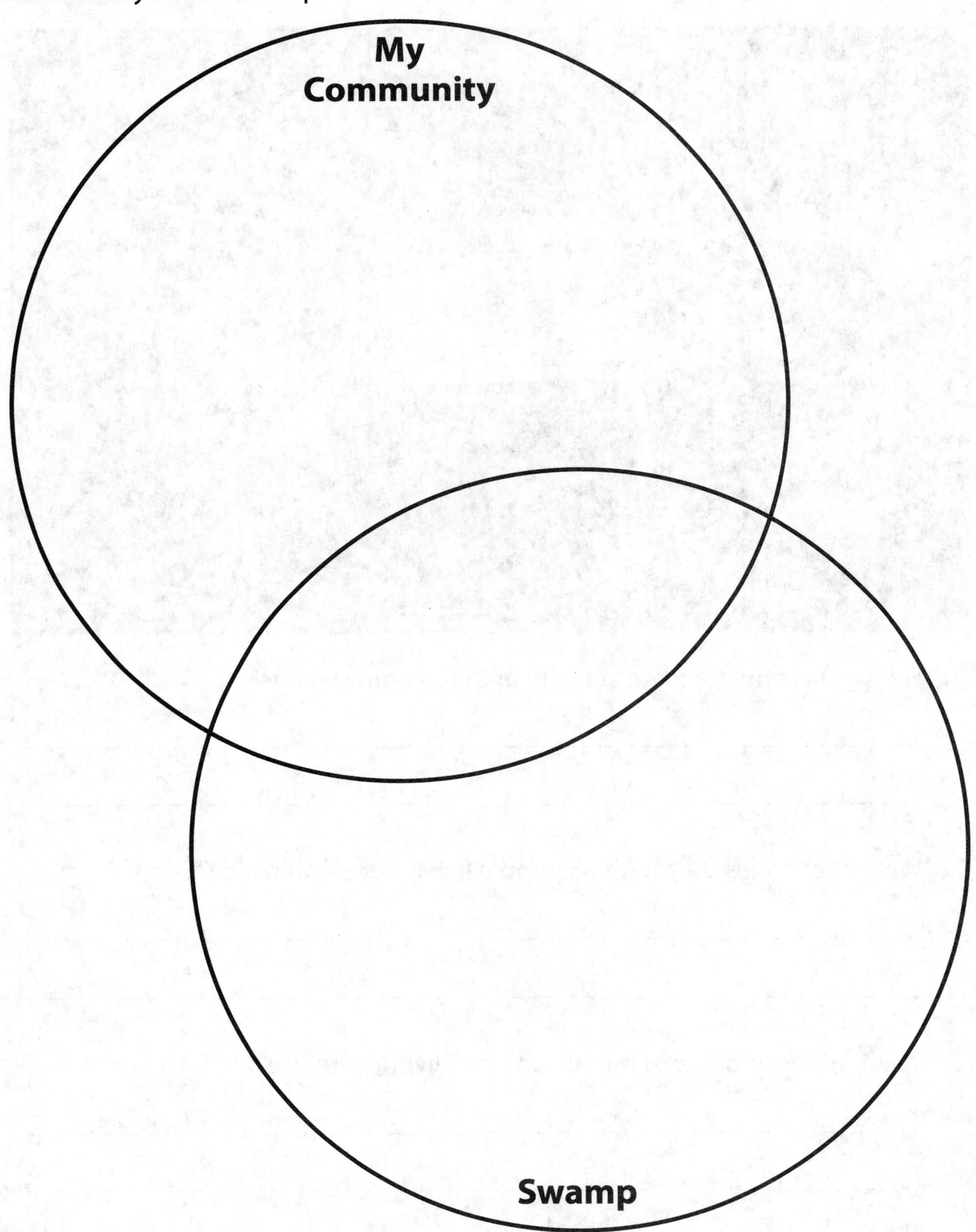

Name: ______________________________ **Date:** ______________

Directions: Study the map, and answer the questions.

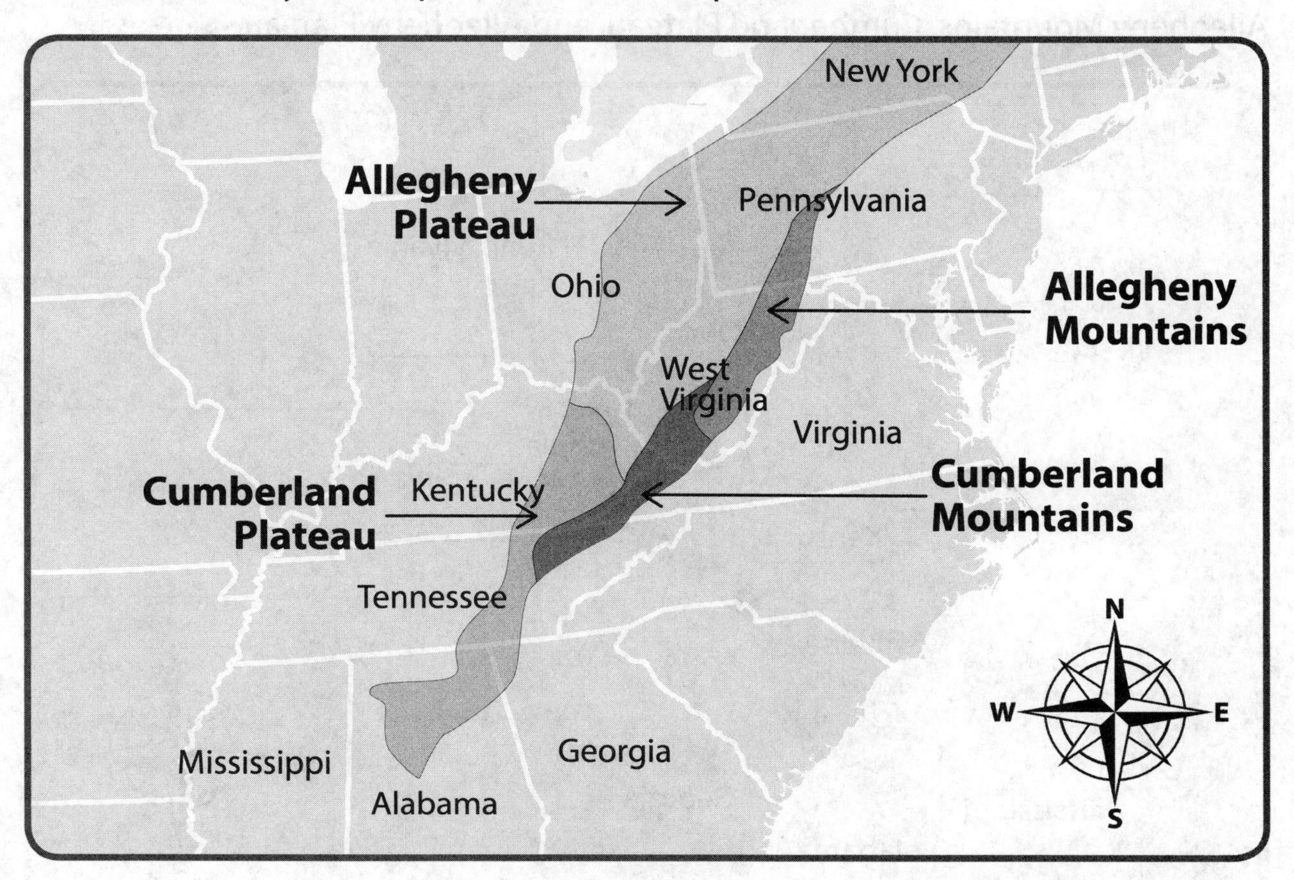

1. In which states is the Cumberland Plateau?

____

____

2. Is more or less than half of the Cumberland Plateau in Kentucky? Explain your thinking.

____

____

3. Which mountain range is east of the Cumberland Plateau?

____

Name: ______________________________ **Date:** ____________________

Directions: Use the clues in the box to label the Cumberland Mountains, Allegheny Mountains, Cumberland Plateau, and Allegheny Plateau.

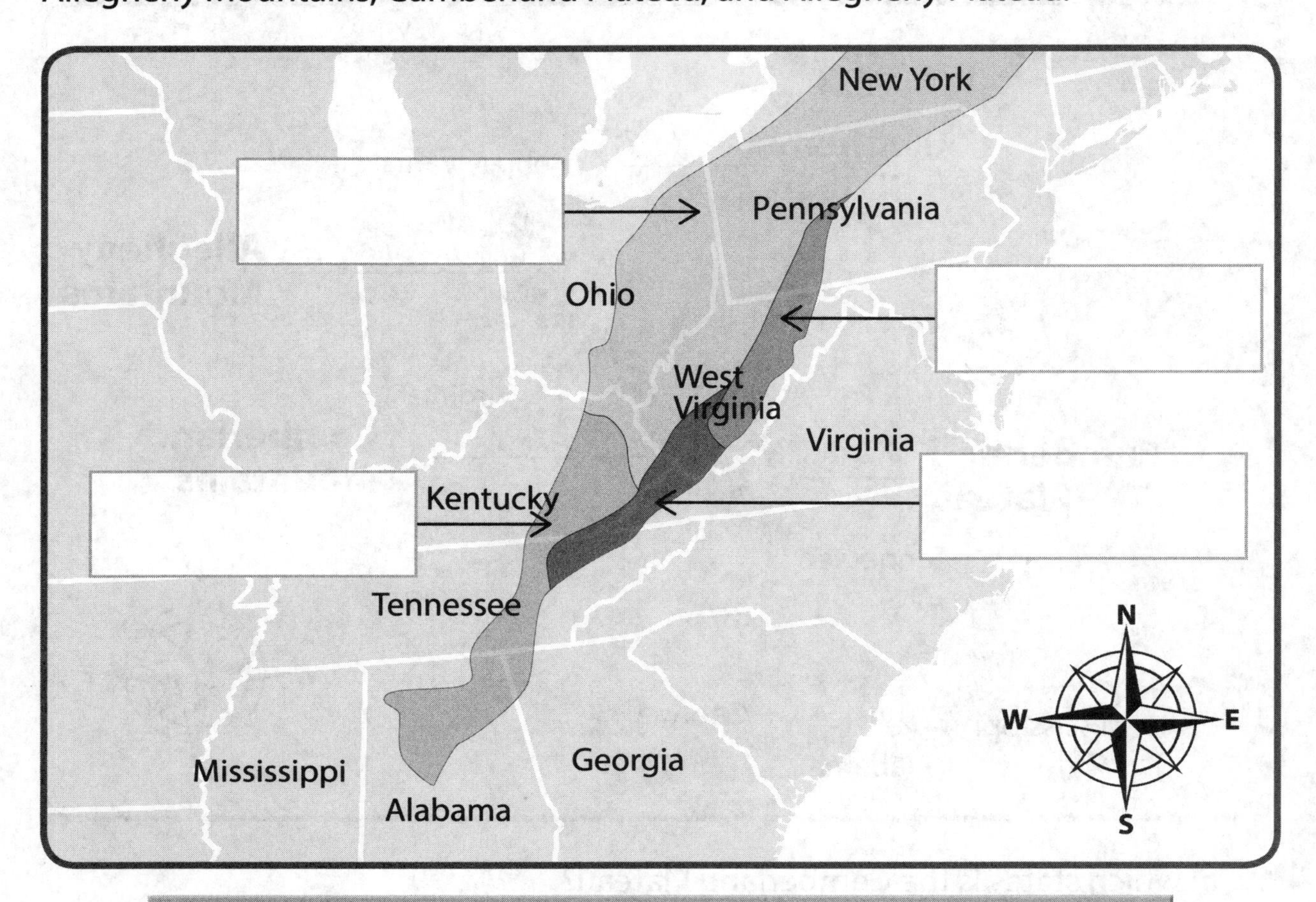

The Cumberland Mountains are in West Virginia, Virginia, Kentucky, and Tennessee.

The Cumberland Plateau is in Alabama, Georgia, Tennessee, and Kentucky.

The Allegheny Plateau is in Ohio, Pennsylvania, New York, West Virginia, and Kentucky.

The Allegheny Mountains extend from West Virginia into Pennsylvania.

Challenge: Cover the names of the states with sticky notes. Write the state names on the sticky notes. Then, check your answers.

Name: ______________________ Date: ______________

Directions: Read the text, and study the photo. Then, answer the questions.

The Cumberland Plateau

The Cumberland Plateau is a raised level stretch of land. Most of the plateau is in Kentucky and Tennessee. It also extends into Georgia and Alabama.

Early people found the land difficult to access. The steep and rocky sides were hard to climb. Early settlers and American Indians lived at the bottom of the plateau.

In the 1900s, people began exploring the plateau. They found that it had many resources. Towns were quickly built on the plateau. Settlers mined coal. They cut trees and set up lumber yards. By the 1950s, most of these resources were gone. Many towns became ghost towns. Homes and businesses were abandoned. People moved to other places.

Today, the Cumberland Plateau is used for recreation. People hike, swim, and explore. People enjoy visiting the many state parks in the region.

1. Why did early people live at the bottom of the plateau?

____

____

2. What is a ghost town? What caused the ghost towns on the Cumberland Plateau?

____

____

3. How is the Cumberland Plateau used today?

____

____

Think About It

Name: ______________________________ **Date:** ________________

Directions: A plateau is flat, elevated land that rises sharply above an area on at least one side. Study the diagram below. Then, answer the questions.

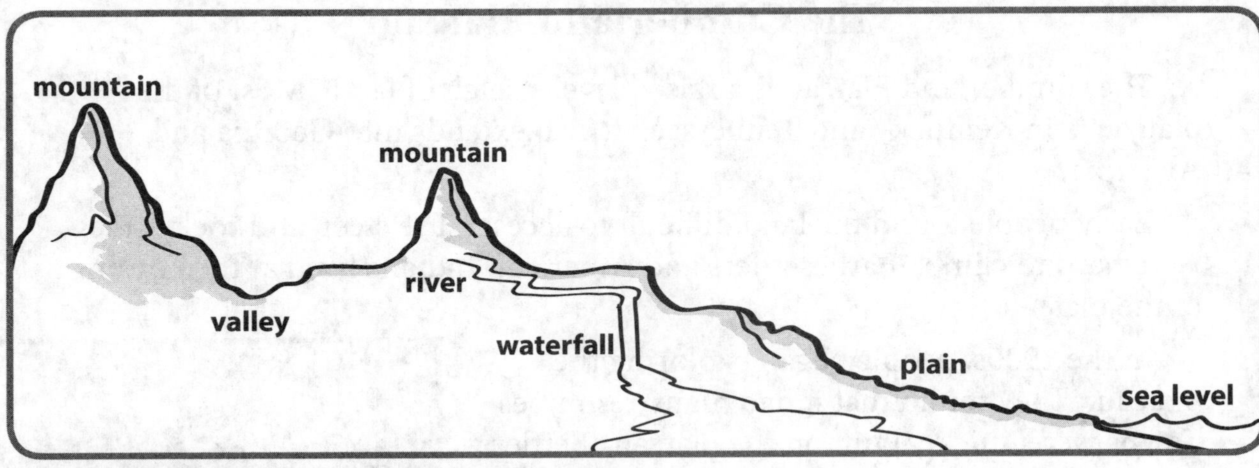

1. Label the two plateaus on this diagram.

2. Describe the locations of the plateaus.

____

____

____

3. What is the difference between a valley and a plateau?

____

____

____

4. How does the water from the river form a waterfall?

____

____

____

Name: ______________________________ Date: ______________

Directions: Draw a plateau next to a mountain. Draw you and your family on top of the plateau. Then, answer the questions.

Geography and Me

1. How would your family get to the top of the plateau?

____

____

2. What are some things you might see on the plateau?

____

____

3. What are some things you and your family could do on top of the plateau?

____

____

Reading Maps

Name: ______________________ Date: ______________

Directions: Study the map of northern Illinois, and answer the questions.

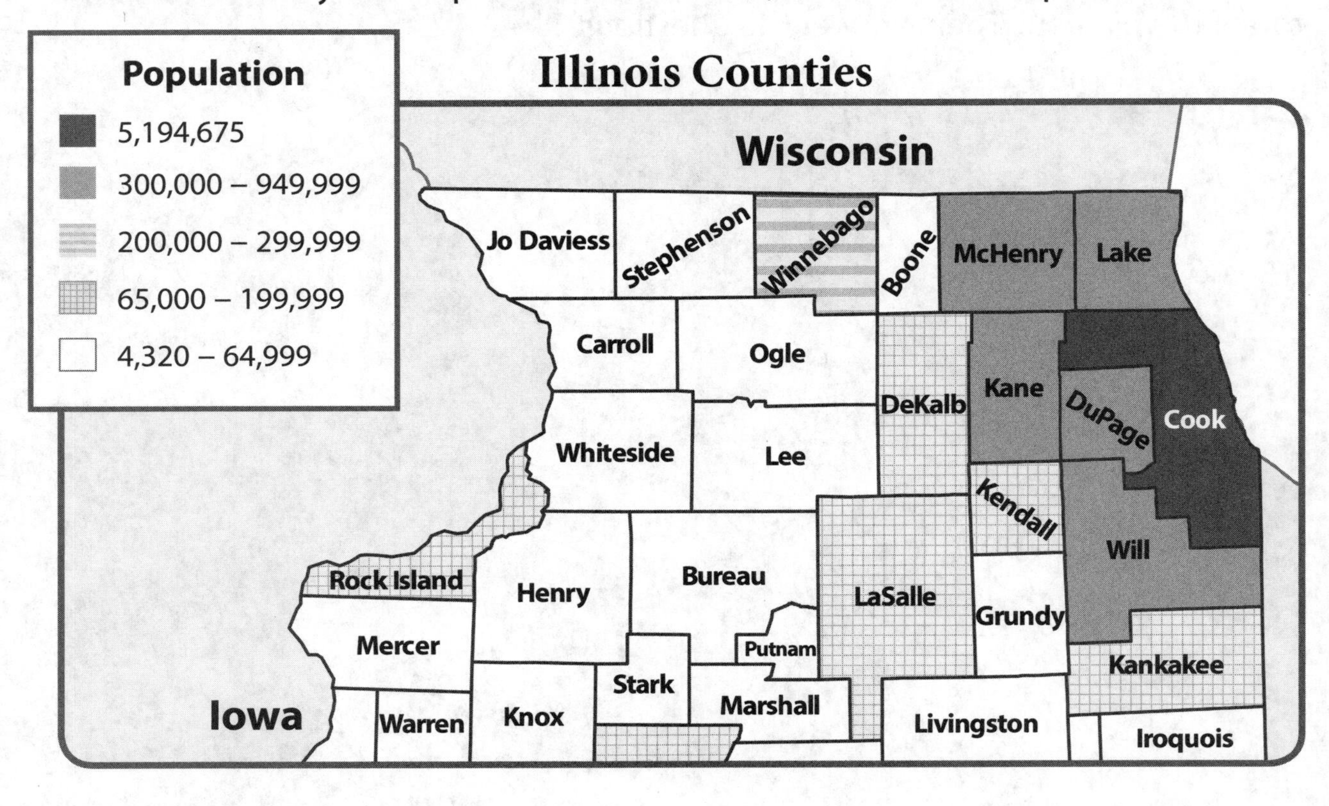

1. Which county has the greatest population?

____

2. Describe the populations of the counties surrounding the county in question 1.

____

____

3. Name a county that borders Marshall County with a greater population.

____

4. Is it reasonable to say that Lake County has 123,456 people? How do you know?

____

____

Name: ______________________________ Date: ____________________

Directions: Locate and label the following cities in each county.

Illinois Counties

City	County
Chicago	Cook County
Aurora	Kane County
Rockford City	Winnebago County
Bloomington	McLean County
Quincy City	Adams County

Read About It

Name: ______________________ **Date:** ______________

Directions: Read the text, and answer the questions.

Chicago Metropolitan Area

The city of Chicago is in northeastern Illinois. It is one of the largest cities in the United States. Nearly three million people live there. It is part of a larger metropolitan area. This area includes the city as well as the towns and suburbs outside the city.

Many people enjoy living in the city. Streets and buildings are close to one another. Stores and restaurants are within walking distance from home. People like living and working in tall skyscrapers.

Others think the city is too crowded. They may work in the city but choose not to live there. Instead, they commute from places known as *suburbs*. Suburbs are smaller neighborhoods located outside the city. They are usually quieter than the city. There is less traffic and fewer tall buildings. Homes are more spread out in the suburbs.

1. What is a metropolitan area?

____

____

2. What are the advantages of living in a suburb?

____

____

____

3. What are the advantages of living in a big city?

____

____

____

Name: ______________________ Date: ____________

Directions: This is a photo of Chicago. Study the photo, and answer the questions.

Think About It

1. Describe the buildings in the photo.

____

____

2. What types of pollution do you think are found in cities such as Chicago?

____

____

3. How has this land been affected by the streets and buildings in this photo?

____

____

____

4. Willis Tower is the tallest skyscraper in Chicago. It is 1,450 feet. Circle it.

Geography and Me

Name: ______________________________ Date: ____________________

Directions: Imagine that you and your family are moving to a new area. Would you prefer to move to a big city or to a suburb? Provide at least two reasons to defend your choice.

Name: ______________________________ Date: ______________

Directions: Ohio and Indiana are home to many types of tourist attractions. These include museums, sports stadiums, amusement parks, and zoos. Study the map of the zoo, and answer the questions.

State Zoo

1. You enter the zoo and head east on the path. Which two animal exhibits will you first see on your right and left?

____

2. A visitor enters the zoo and walks north to the monkeys. Then, she walks west near the horses and around back to the entrance. How many exhibits did she see?

____

3. Use cardinal directions to describe one way to get from the entrance to the elephant exhibit.

____

____

____

Name: ______________________________ **Date:** ________________

Directions: The Ramirez family is excited to see some of the exhibits at the zoo. Follow the steps to draw a line that shows the path they take in the zoo.

1. The family enters the zoo and heads north.

2. They stop at the turtle exhibit.

3. From there, the family heads east past the food stand. They keep walking until they get to the cheetah exhibit.

4. Then, they walk south and spend time watching the bears.

5. Next, the family turns around and walks north and west until they reach the horses.

6. Finally, they walk south to leave the zoo.

Name: ______________________________ Date: ________________

Directions: Read the text, and study the photo. Then, answer the questions.

Tourist Attractions

People who visit other places to see the sights are called *tourists*. Indiana and Ohio are home to many fun tourist attractions.

Indiana is known for its many sports teams. People come from all over to see basketball and football games. The Indianapolis 500 is a world-famous car race. Drivers race around a track 500 times. Thousands of people come to see this race.

Ohio also has many unique attractions. The Rock and Roll Hall of Fame and Pro Football Hall of Fame are in this state. Amusement parks bring people to the state, too. Tourists visit Cedar Point and King's Island in the summer. They enjoy the large coasters and thrill rides.

Both states also have museums and zoos. People visit national parks and forests. All these attractions help the economies of the states.

1. What is a tourist?

____

____

2. What are some tourist attractions in Ohio?

____

____

3. Why do you think the Indianapolis 500 is popular with tourists?

____

____

Think About It

Name: ______________________________ Date: ________________

Directions: Indiana and Ohio are each home to sports teams. Some teams play in large stadiums. There are many jobs workers must do before, during, and after games. Study the photo of the stadium, and answer the questions.

Victory Field in Indiana

1. What jobs might workers do before a sporting event?

____

____

____

2. What jobs might workers do during a sporting event?

____

____

____

3. What jobs might workers do after a sporting event?

____

____

____

Name: ______________________________ Date: ________________

Directions: What is a popular tourist attraction near you? Create an advertisement about this place. Include a drawing and words to attract people to this place.

Reading Maps

Name: ______________________ **Date:** ______________

Directions: Lewis and Clark were explorers in the early 1800s. The Lewis and Clark National Historic Trail lets people see the route these explorers took. This map shows part of the trail. Study the map, and answer the questions.

Lewis and Clark National Historic Trail

Population
points of interests
capital
Lewis and Clark Trail
Three Affiliated Tribes Museum
Lewis and Clark Interpretive Center
Bismarck
North Dakota
South Dakota
Oahe Visitor Center
Pierre
Akta Lakota Museum
Lewis and Clark Information Center
N
S
E
W

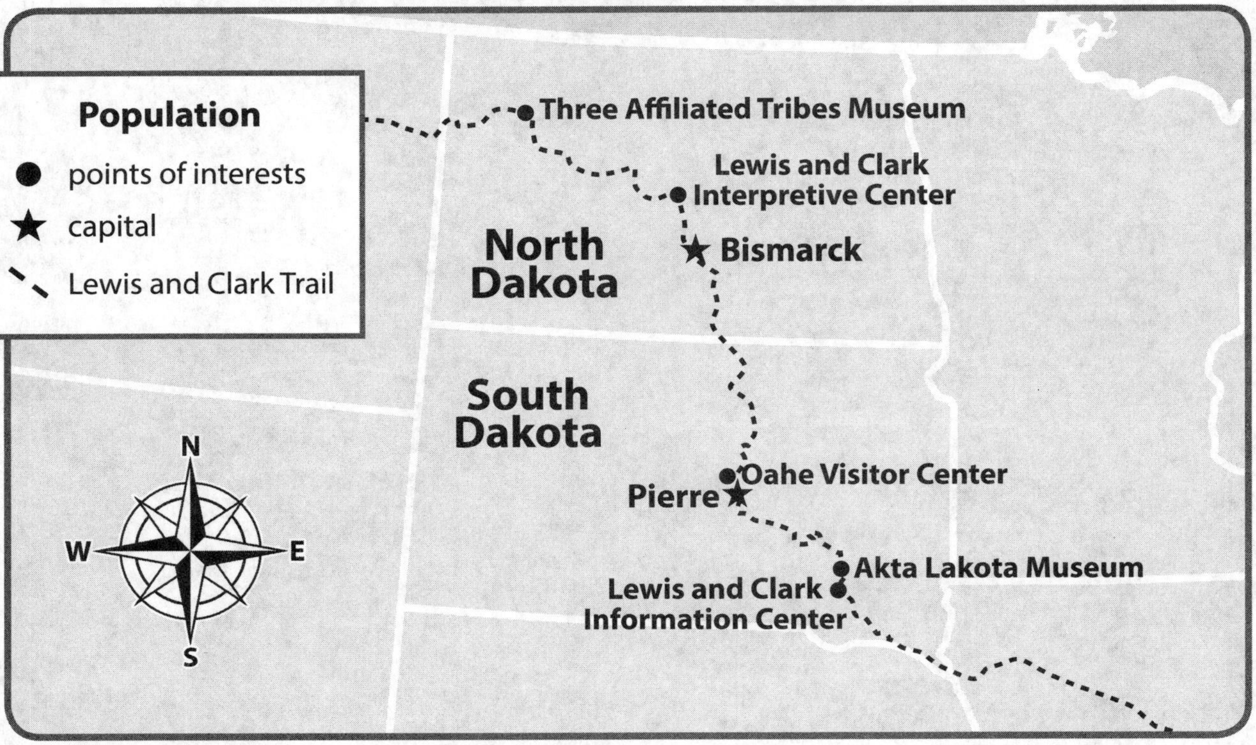

1. What two museums are on this map?

____

____

2. What are the capitals of North Dakota and South Dakota? How do you know?

____

____

3. Describe the route of the trail through these two states.

____

____

Name: ______________________________ Date: ________________

Directions: Points of interests are places on a map that people might like to see. Create a map of your school or neighborhood. Include points of interest on your map. Include a legend to help readers understand your map.

Title: ______________________________

Read About It

Name: ______________________ **Date:** ______________

Directions: Read the text, and study the photo. Then, answer the questions.

Mount Rushmore

Doane Robinson came up with the idea for Mount Rushmore. He wanted to attract visitors to South Dakota. He wanted to carve the faces of four presidents into the side of a mountain. The project took 14 years to complete.

The monument was to be a symbol of the nation. The choice of presidents was important. George Washington was the first president and a great leader. Thomas Jefferson was also chosen. He also helped found the country. Theodore Roosevelt was chosen, too. He was a great leader. He helped the country grow. Abraham Lincoln is also carved into the mountain. He helped keep the country together during the Civil War. He also helped end slavery.

Today, millions of people visit Mount Rushmore. It is a symbol of freedom. It is a tribute to four great presidents.

1. Which presidents are on Mount Rushmore?

____

____

2. Why is Mount Rushmore an important monument?

____

____

____

3. Why might someone want to visit Mount Rushmore?

____

____

Name: ______________________ Date: ______________

Directions: Theodore Roosevelt National Park is in North Dakota. Visitors can stay at the Cottonwood Campground. Study the map of the campground and answer the questions.

Cottonwood Campground

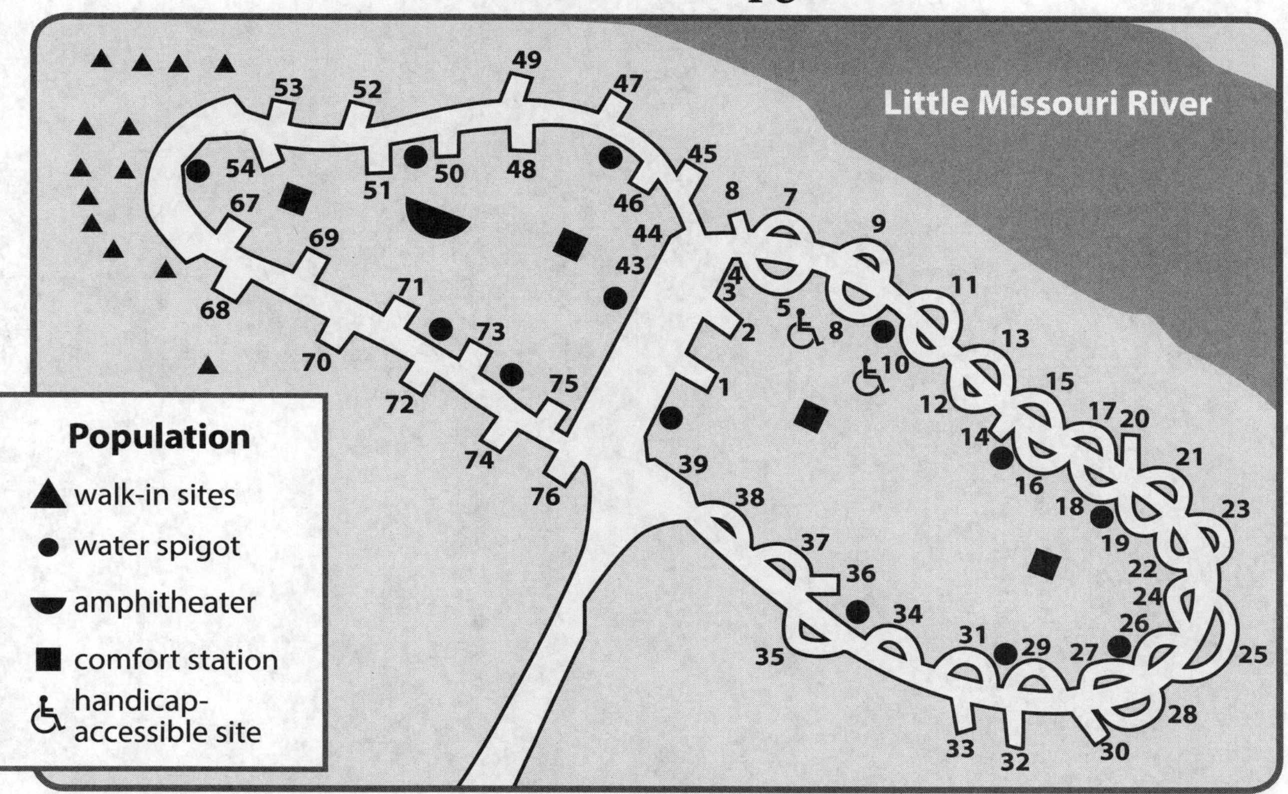

1. Which sites are meant for people with disabilities? How do you know?

____

____

2. How many water spigots are in this campground?

____

3. What body of water is this campground near? What might people do for fun there?

____

____

Name: ______________________ Date: ______________

Directions: Create your own Mount Rushmore. Draw the faces of four important people in your life. Make it look like their faces are carved into a mountain. Then, describe why they deserve to be on the monument.

Name: ______________________________ **Date:** ______________

Directions: Missouri has a lot of erosion. Erosion occurs when wind or water remove rocks or soil from the land. Study the pictures, and answer the questions.

Erosion: Before and After

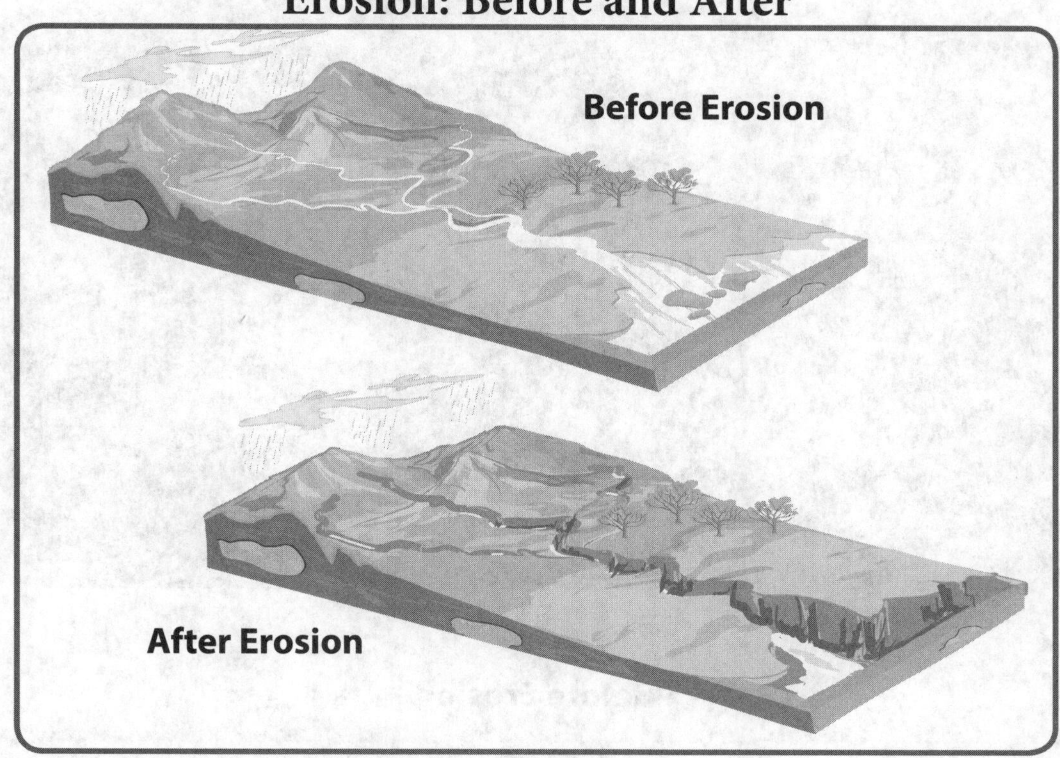

1. How does erosion affect the land?

____

____

2. What changed the most in the pictures above? Describe how it changed.

____

____

3. Do you think rain or rivers causes more erosion? Explain your reasoning.

____

____

Creating Maps

Name: ______________________________ Date: ____________________

Directions: This photo shows land after erosion has taken place. What do you think the land looked like before? Draw it in the space below.

Before Erosion

Name: ______________________ Date: ______________

Directions: Read the text, and study the photo. Then, answer the questions.

Erosion Control

Erosion occurs when wind and water move dirt and rocks from the land. Not all erosion is bad. But sometimes, people need to control it.

Farmers in Missouri are concerned about soil erosion. They rely on topsoil for their crops. Topsoil has nutrients that help plants grow. Wind and water can easily erode topsoil. Farmers must think of ways to prevent this.

One way is to install a windbreak. A windbreak is a row of tall trees or bushes. It helps block the wind from blowing away the soil. Another way farmers prevent erosion is by making a barrier around the soil. They do this by planting crops and other plants around the edge of the soil. The roots from the plants help hold the soil together. Farmers can also prevent erosion by careful watering. All crops need water. But, over-watering can cause soil erosion. Farmers are careful to only water certain parts of the land at once.

People work together to prevent too much erosion. They hope to maintain the land so that people can enjoy everything Missouri has to offer.

1. What is a windbreak?

____

____

2. How do plant roots help prevent erosion?

____

____

3. How might a farmer have prevented the soil erosion in the photo?

____

____

Name: ______________________ Date: ______________

Think About It

Directions: This map shows some of the river basins in the United States. A river basin includes all the land that drains into into a river. Study the map. Then, answer the questions.

River Basins

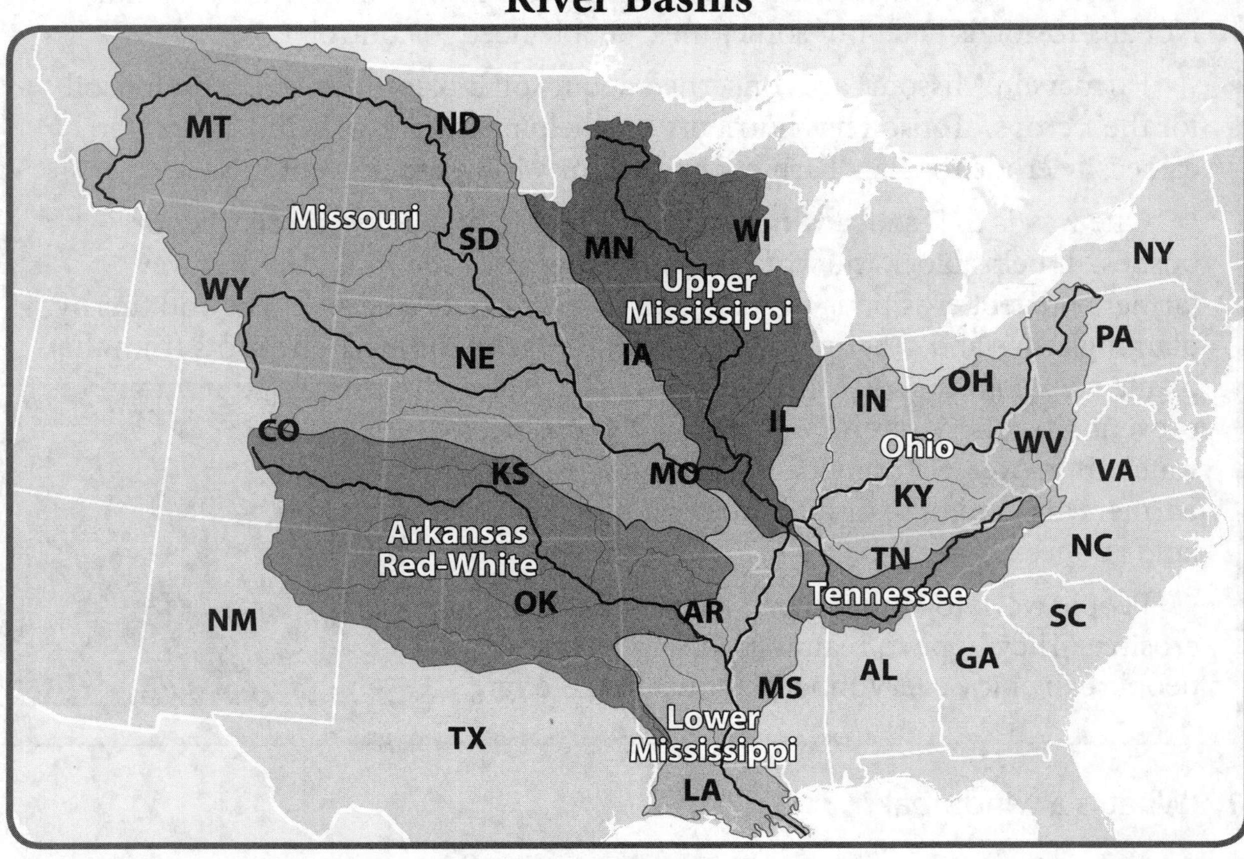

1. Which river basins are in Missouri (MO)?

____

____

____

2. Why might Missouri have so many erosion problems?

____

____

____

Name: ______________________________ Date: ____________________

Directions: What are some areas in your community that could be affected by erosion? Draw and label two places.

Reading Maps

Name: ______________________ **Date:** ______________

Directions: Study the map of Kansas. Then, answer the questions.

Kansas Land Use

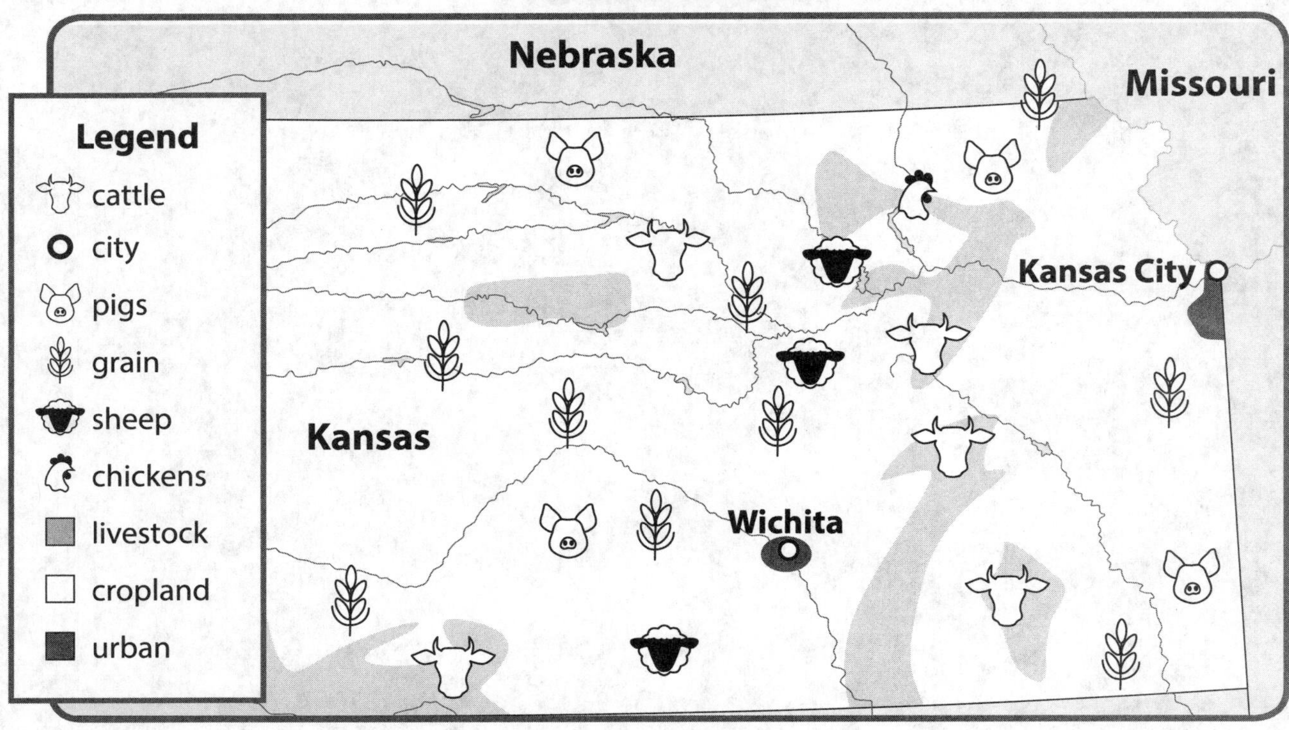

1. What is most of the land in Kansas used for?

____

2. Based on the map, what do Kansas City and Wichita have in common?

____

____

3. What livestock, or farm animals, are represented on this map?

____

____

Challenge: Cover the names of the states with sticky notes. Write the state names on the sticky notes. Then, check your answers.

Name: ______________________________ Date: ________________

Directions: Follow the steps to complete the map.

Nebraska Land Use

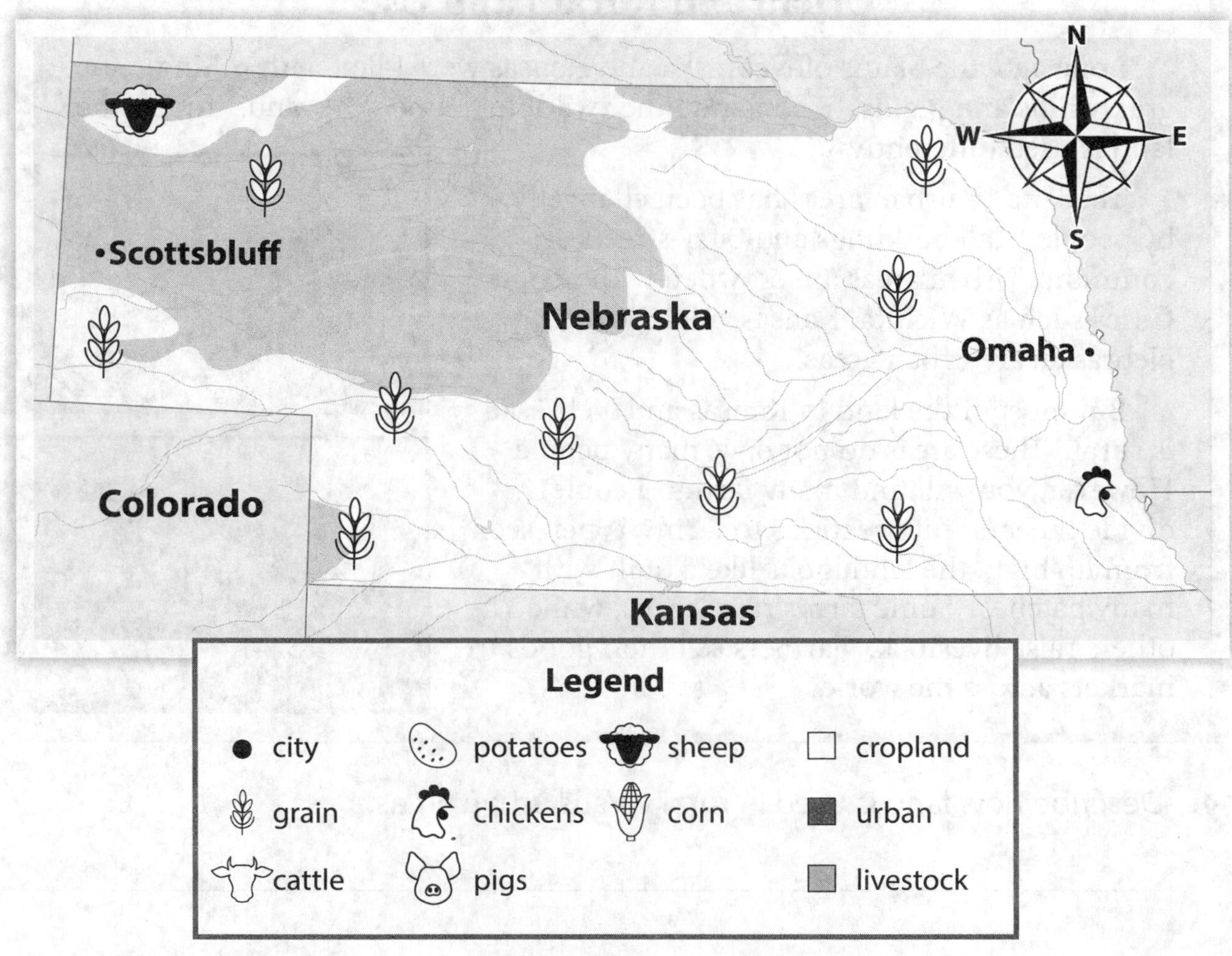

Legend

city	potatoes	sheep	cropland
grain	chickens	corn	urban
cattle	pigs		livestock

1. Add the symbol for potatoes near Scottsbluff.
2. Shade a circle around Omaha to show this as an urbanized area.
3. Add symbols for cattle, pigs, and corn near the southeastern corner of the state.

Challenge: Cover the names of the states with sticky notes. Write the state names on the sticky notes. Then, check your answers.

Read About It

Name: ______________________ Date: ______________

Directions: Read the text, and study the photos. Then, answer the questions.

Urban and Rural Land Use

Long ago, the plains of Nebraska and Kansas were filled with rolling, grassy hills and plains. Bison and other wildlife roamed the land. Today, the land is used differently.

The land in urban areas has been changed by people. Tall buildings and busy streets are common. Urban areas are crowded with people. Cities such as Wichita, Kansas, and Omaha, Nebraska, are urban areas.

But most of the land in Kansas and Nebraska is rural. These areas do not have many people. However, you will find many farms. People divide the land into sections to farm. When seen from up high, the land looks like a quilt with many patches. Some farms raise crops, while others raise livestock. Farmers sell their goods to markets across the world.

1. Describe how land is used in rural versus urban areas.

____

____

____

2. How have people changed the land in each photo?

____

____

____

____

Name: ______________________________ **Date:** ______________

Directions: Many people in the United States live in urban areas. Study the map, and answer the questions.

Populations of Big Cities

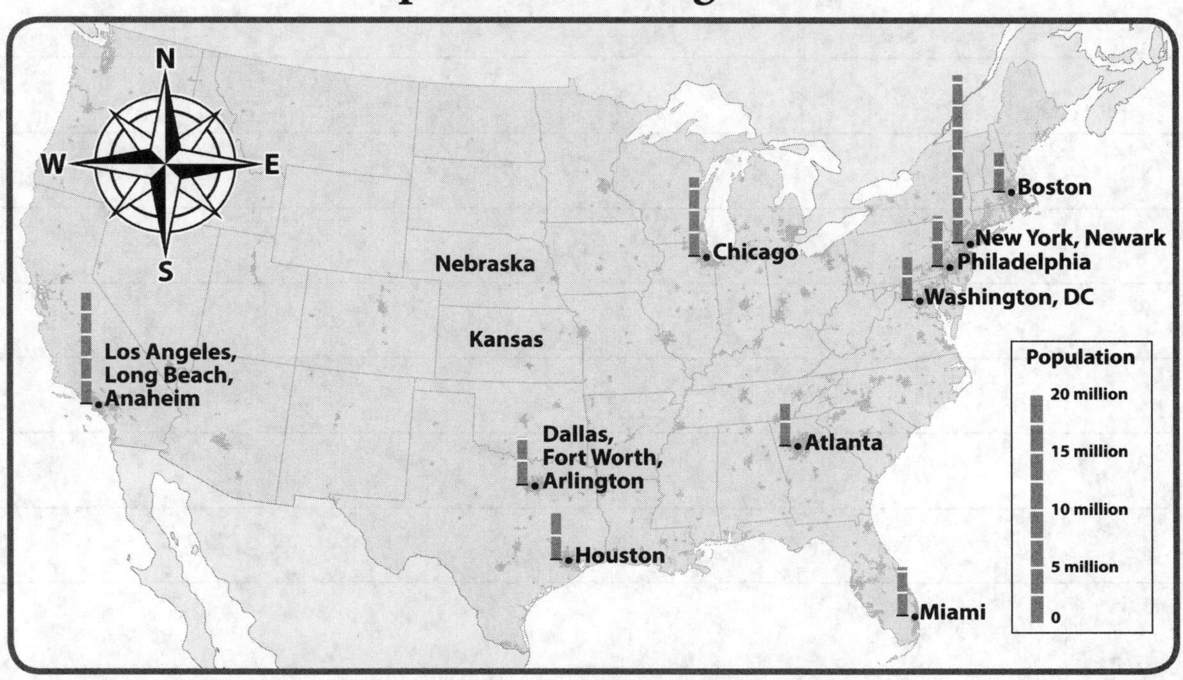

1. Do more people live in the eastern or western half of the United States? Explain your thinking.

____

____

2. About how many people live in the Atlanta and Miami areas combined? How do you know?

____

____

3. How would this map be different if it showed rural areas rather than urban areas?

____

____

Name: ______________________________ Date: ______________

Geography and Me

Directions: How might the land in your community have been changed by people? List the ways people may have changed the land. Then, draw what you think this area may have looked like long ago.

____

____

____

____

____

____

____

____

Name: ______________________ Date: ______________

Directions: This is a grid map of a farm. Study the map, and answer the questions.

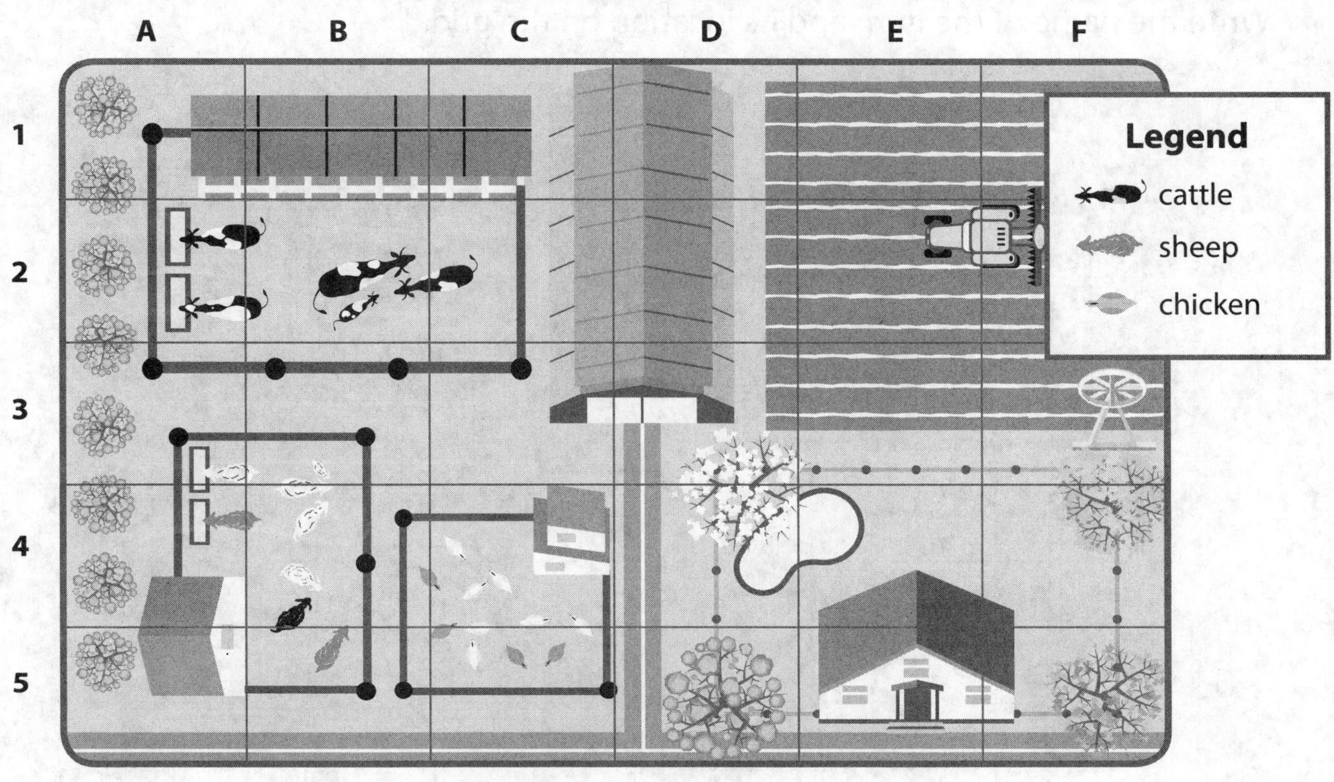

Reading Maps

1. Use the grid to describe the location of the field.

____

2. Use the grid to describe the location of the sheep.

____

3. What is located at D5?

____

4. What is located at A2?

____

5. What is located at E5?

____

Name: ______________________________ Date: ________________

Directions: Draw five more items on the map that might be found on a farm. Write the name of the item and its location on the grid.

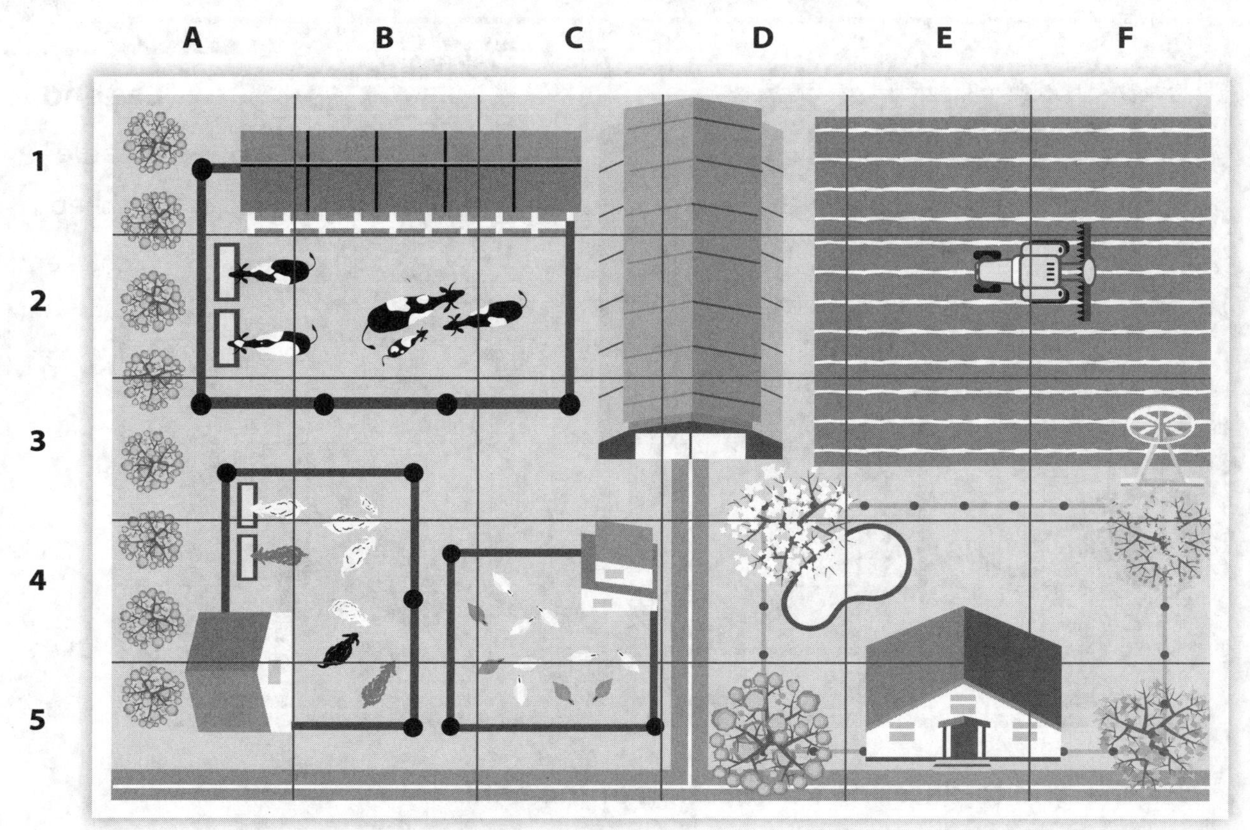

1. ______________________________

2. ______________________________

3. ______________________________

4. ______________________________

5. ______________________________

Name: ____________________ Date: ____________

Directions: Read the text, and answer the questions.

Agriculture

Farming is a big business in Iowa and Wisconsin. Many farmers grow and sell food. They also raise and sell animals. This type of industry is known as *agriculture*.

Iowa is one of the largest producers of corn in the country. Soybeans and apples are also grown there.

Wisconsin has many dairy farms. Dairy products are made from milk. Wisconsin is known for its cheese, which is a dairy product.

Many farms in these states are very large. It would be difficult to plant, water, and maintain these crops by hand. There are many tools to help farmers grow their crops. Tractors are used to pull things around the farm. A plow is a machine that helps turn the soil over. This loosens the dirt to make it better for planting. An irrigator machine spreads water quickly through the farm. There are even machines to help plant seeds and harvest crops. Each machine helps farmers save time and money. That lets farmers get their food to more people.

1. What main crops are grown in Iowa?

____

____

2. What does a plow do?

____

____

3. How do tools help farmers grow more food?

____

____

Think About It

Name: ______________________ **Date:** ______________

Directions: This table shows dairy products from Wisconsin. Study the table, and answer the questions.

Product	Wisconsin Production	Rank (compared to other states)
cheddar cheese	613,467, 000 lbs	1
mozzarella cheese	1,051,356,000 lbs.	2
provolone cheese	200,982,000 lbs.	1
parmesan cheese	127,135,000 lbs.	1
romano cheese	29,364,000 lbs.	1
limburger cheese	481,000 lbs.	1
milk	29,030,000,000 gal.	2

1. How many of these Wisconsin products are ranked number one?

____

2. Is the combined amount of romano and parmesan cheeses produced more than 200 million pounds? How do you know?

____

____

3. Based on the chart, describe how Wisconsin's cheese production compares to other states.

____

____

4. Why might Wisconsin be known as the Dairy Capital of the United States?

____

____

Name: ______________________________ Date: ____________________

Directions: Farming on a large field requires large tools. What tools could you use to make gardening easier? Write and draw one tool that would help you do each task.

Task	Tool
loosening the soil	
watering	
moving dirt	
pulling objects around	

Reading Maps

Name: ______________________ Date: ______________

Directions: This is a map of the Great Lakes. Study the map, and answer the questions.

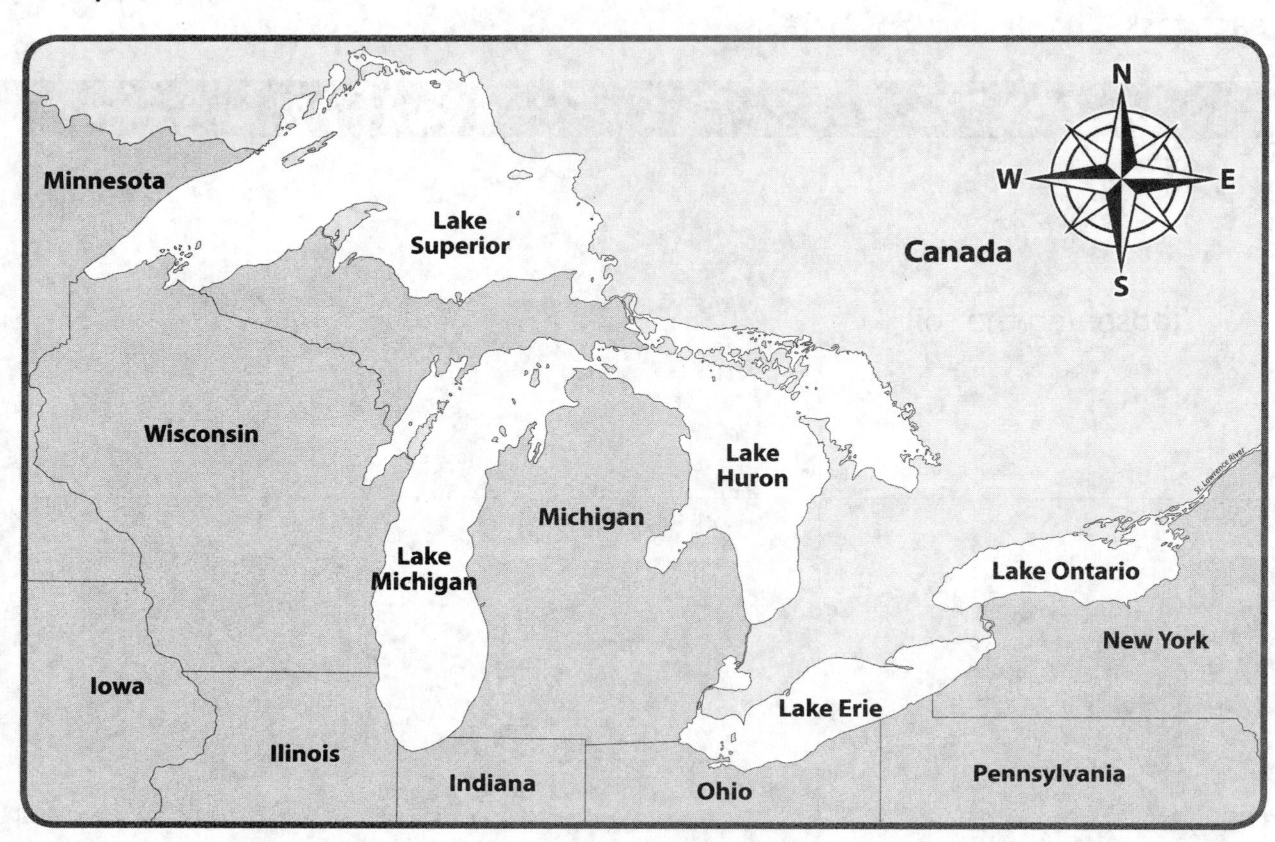

1. Which of the Great Lakes border Wisconsin?

____

____

2. The names of the Great Lakes can be remembered using the acronym HOMES. Complete the acronym below.

H ____

O ____

M ____

E ____

S ____

Name: ______________________ Date: ______________

Directions: Use the clues to label each of the Great Lakes on the map.

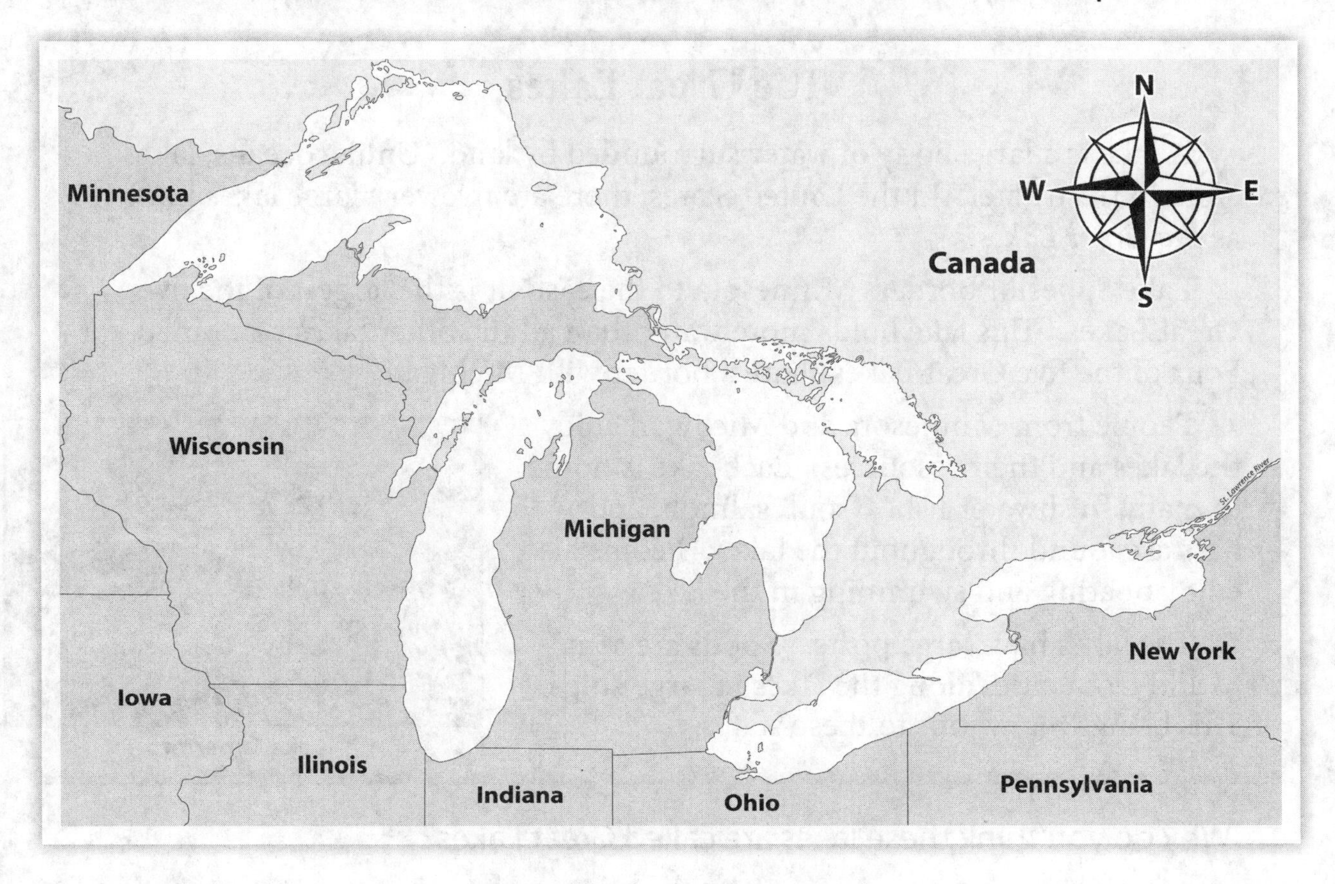

1. Lake Huron borders Michigan and Canada.
2. Lake Ontario is the farthest east.
3. Lake Michigan splits the state of Michigan in half.
4. Lake Erie borders four states.
5. Lake Superior is the largest lake.

Challenge: Cover the names of the states with sticky notes. Write the state names on the sticky notes. Then, check your answers.

Read About It

Name: ______________________ **Date:** ______________

Directions: Read the text, and study the photo. Then, answer the questions.

The Great Lakes

A lake is a large body of water surrounded by land. Unlike oceans, lakes contain freshwater. In the United States, there are five very large lakes known as *the Great Lakes.*

Lake Superior borders Minnesota to the east. It is the largest of the five Great Lakes. This lake holds more water than all the other lakes combined. Four of the five Great Lakes share a border with Michigan.

People from Minnesota and Michigan enjoy the lakes and their resources. Each lake is home to many freshwater fish. Trout, salmon, and bass are found throughout the lakes. People also enjoy boating and swimming in the lakes.

The lakes have large ports. Goods are sent to and from cities along the lakes in large ships. This brings many jobs to these states.

Lake Superior

1. Why do you think these lakes are called Great Lakes?

____

____

2. Why might someone visit the Great Lakes?

____

____

3. How is the lake in the photo similar to and different from an ocean?

____

____

____

Name: ______________________ Date: ______________

Directions: This graph shows how many jobs are connected to the Great Lakes. Study the graph, and answer the questions.

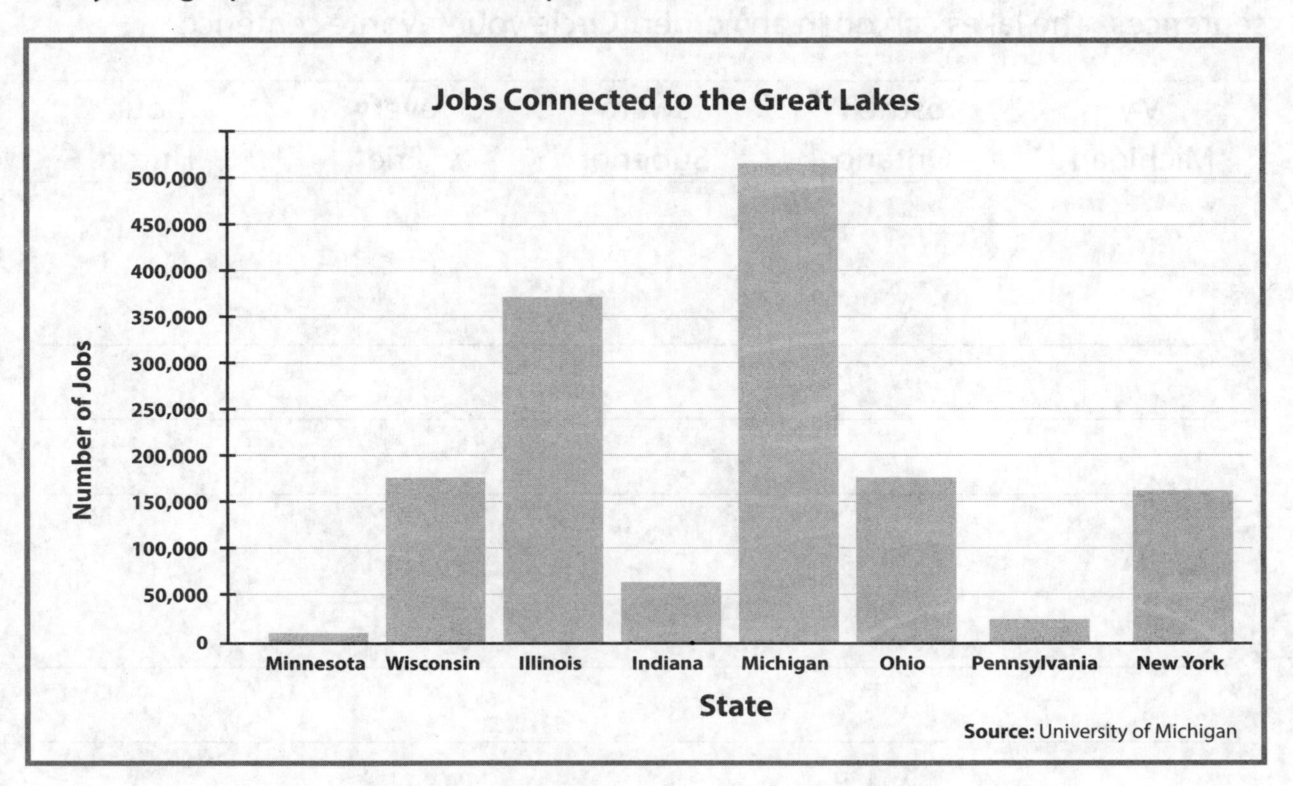

1. True or False? The number of jobs produced in Ohio is more than twice the amount in Indiana. Explain your reasoning.

____

____

____

2. About how many jobs are produced in Wisconsin and Illinois combined?

____

3. Why do you think Michigan has the most Great Lakes jobs?

____

____

Name: ______________________ **Date:** ______________

Directions: Some people make up silly sentences to help them remember the names of the Great Lakes. Look at the example in the box. Then, write your own sentences. The lakes can go in any order. Circle your favorite sentence.

My	ostrich	swam	every	hour.
Michigan	Ontario	Superior	Erie	Huron

1. ______________________________

2. ______________________________

3. ______________________________

4. ______________________________

Name: ______________________ Date: ______________

Directions: This is a weather map for part of Oklahoma. Study the map, and answer the questions.

Weather in Oklahoma

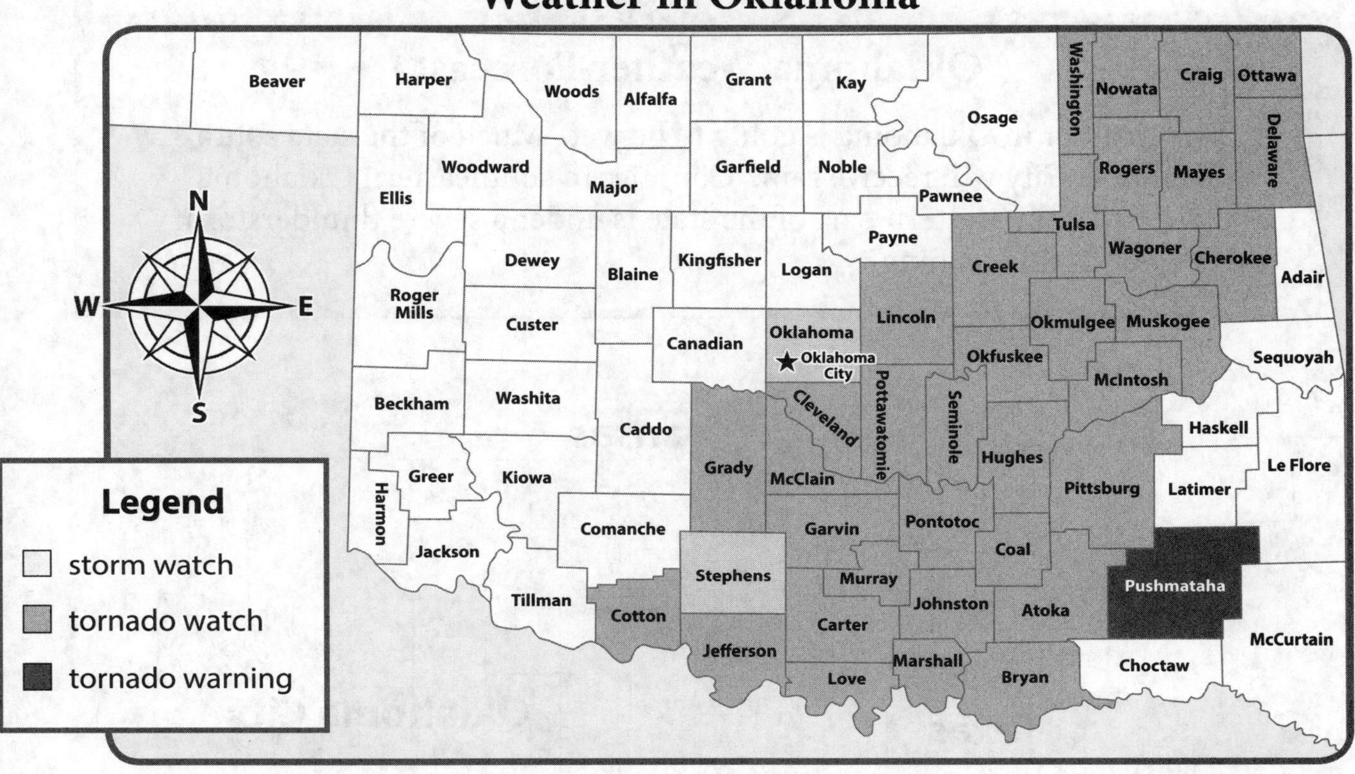

1. Which county has had a tornado warning issued?

____

2. What type of weather is expected in Oklahoma City?

____

3. What part of the state has a tornado watch or warning?

____

4. Is a tornado watch or a tornado warning more severe? How do you know?

____

____

Name: ______________________ **Date:** ______________

Directions: Read the weather forecast in the box. Then, color the map to show the weather that is predicted. Create a legend to help readers understand your map.

Oklahoma Weather Forecast

The weather in Oklahoma is going to be wet. Much of the state south of Oklahoma City will receive rain. Counties in southeastern Oklahoma may see hail. The western part of the state is under a severe thunderstorm warning until this evening.

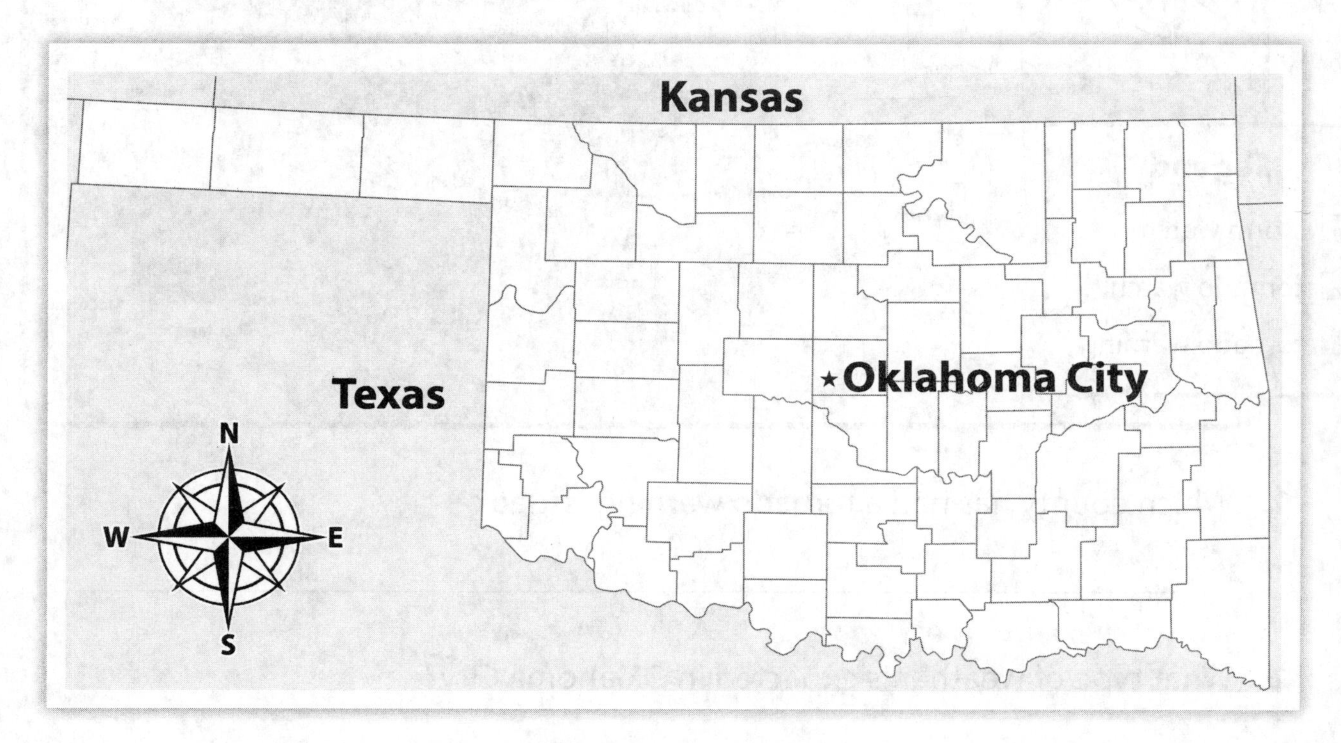

Legend

Name: ______________________________ Date: ______________

Directions: Read the text, and study the photo. Then, answer the questions.

Tornados

Oklahoma is one of many states in a region known as *Tornado Alley*. It extends from Texas north through South Dakota. This region receives many of these extreme storms. They form when warm air from the Gulf of Mexico combines with dry air from states such as New Mexico and Arizona. When cool air from Canada is added, it becomes the perfect condition for a tornado.

The National Weather Service tracks this type of weather. They have a system to alert people about the threat of these storms. The first level is a tornado watch. A watch is issued when conditions are right for a tornado. This does not mean there actually is a tornado. A tornado warning means a twister has formed. A warning is also issued when wind high in the sky begins to spin in the clouds. This is known as a *funnel cloud*. Funnel clouds can turn into tornados. Watches and warnings are announced to give people time to prepare.

1. Who decides whether a watch or a warning is announced?

____

2. What is Tornado Alley?

____

____

3. Should a tornado watch or a warning be issued for the town in the photo? Explain your thinking.

____

____

____

Name: ______________________ **Date:** ______________

Directions: Scientists use the Enhanced Fujita Scale to classify tornados. A Fujita rating tells how much damage a tornado can cause. Study the scale below, and answer the questions.

EF-0
wind: 65–85 mph
minor damage

EF-1
wind: 86–110 mph
moderate damage

EF-2
wind: 111–135 mph
considerable damage

EF-3
wind: 136–165 mph
severe damage

EF-4
wind: 166–200 mph
extreme damage

EF-5
wind: > 200 mph
incredible damage

1. At what rating do you begin to notice roof damage?

____

2. How would you describe the destruction caused by an EF-5 tornado?

____

____

3. During a tornado, people are advised to stay in underground basements. Why do you think that is a safe place?

____

____

Name: ______________________________ Date: ______________

Directions: What natural disasters are common where you live? Write and draw what your family can do to prepare.

____

____

____

____

____

____

____

____

____

Reading Maps

Name: ______________________ Date: ______________

Directions: Study the map, and answer the questions.

North American Deserts

1. Which desert is the smallest?

____

2. Which desert is partly in New Mexico?

____

3. What would you expect the climate to be like in this region? Why?

____

____

Name: ______________________ Date: ______________

Directions: Use the clues to label the Mojave, Chihuahuan, Sonoran, and Great Basin deserts on the map.

North American Deserts

1. The only major desert in Texas is the Chihuahuan Desert.

2. The Mojave Desert is south of the Great Basin Desert.

3. The Sonoran Desert is in southwestern Arizona.

4. The Great Basin Desert is the farthest north.

Name: ______________________ **Date:** ____________

Read About It

Directions: Read the text, and study the photo. Then, answer the questions.

The Chihuahuan Desert

The Chihuahuan Desert is found in the Southwest. The temperatures in this region are extreme. Daytime temperatures can reach over 100° F (38°C). At night, temperatures can dip below freezing.

Many plants are unique to this desert. Some species of cacti are found here. These plants have special adaptations to help them survive. The sharp needles on a cactus help keep predators away. A cactus is able to store water for a long time. The shin dagger and ocotillo are also found in this desert. These plants have sharp leaves to protect them.

Animals such as snakes, prairie dogs, foxes, and jackrabbits call this desert home. Some snakes stay cool by burrowing under the soil. They also avoid the heat by hiding in shady areas, such as under rocks. Other animals only come out at night.

1. What adaptations do plants have in the desert?

____

____

2. How do snakes stay cool in the desert?

____

____

3. How might only coming out at night help an animal survive in the desert?

____

____

Name: ______________________________ Date: ______________

Directions: The Pueblo Indians are known for building homes beneath cliffs in Arizona and New Mexico. These are known as *cliff dwellings*. Study the photo of the cliff dwelling, and answer the questions.

1. How might these homes keep cool?

____

____

2. What materials might the building be made from?

____

____

3. What dangers might exist for people living in these homes?

____

____

4. What might be the benefits of living in cliff dwellings?

____

____

Geography and Me

Name: ______________________________ Date: ________________

Directions: Draw a desert scene. Add as many desert plants and animals as you can. Include one animal adaptation in the scene.

Name: ______________________ Date: ______________

Directions: Texas is home to many Spanish missions. A mission is a religious building built in the 17th and 18th centuries. Study the map, and answer the questions.

Missions in Texas

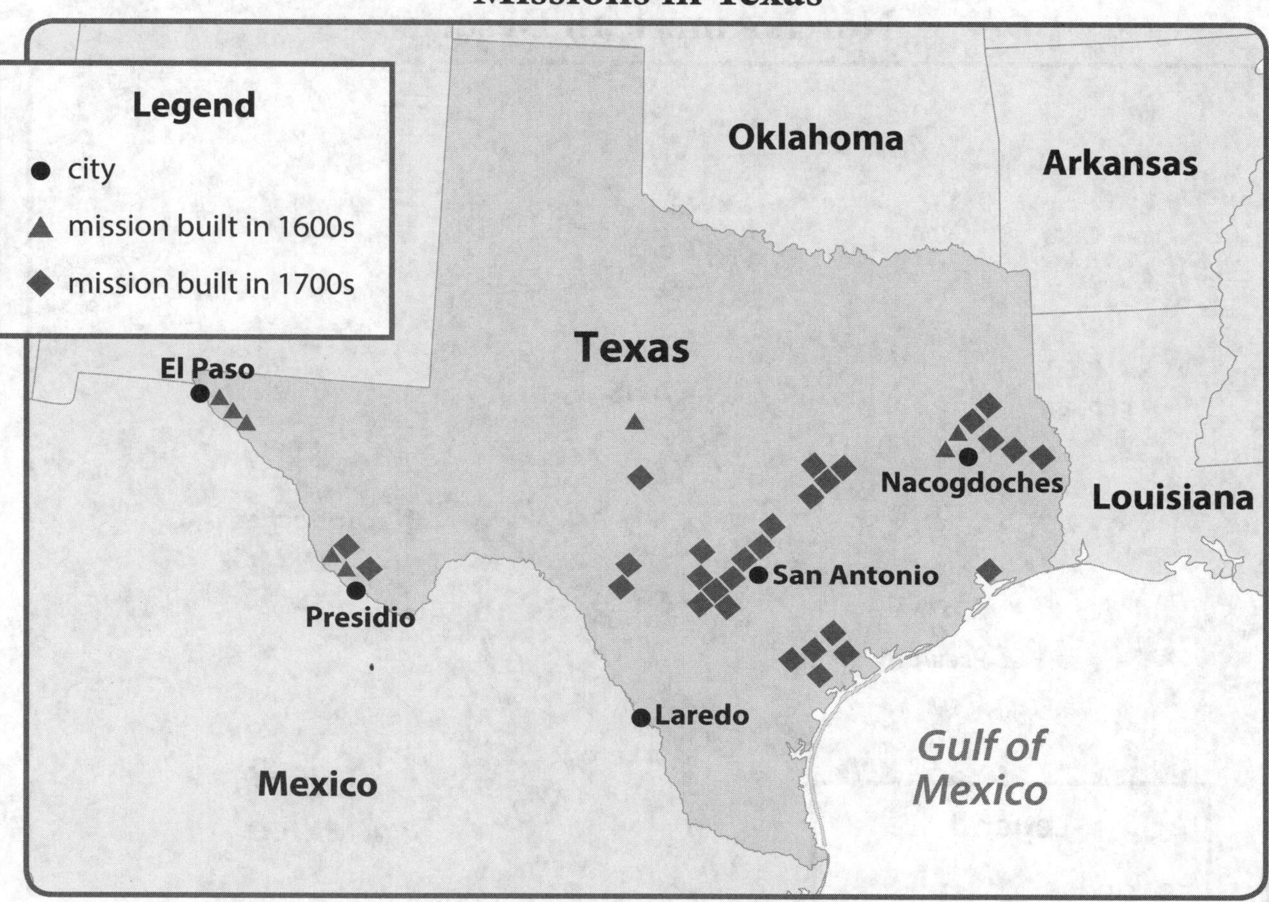

1. How many missions are located near El Paso, Texas?

____

2. How many missions on this map were built in the 1600s?

____

3. What do the presence of these Spanish missions tell you about the state's history?

____

____

Creating Maps

Name: ______________________ Date: ______________

Directions: To protect the missions in Texas, Spain also built presidios, or forts. Create a symbol for the presidios, and add it to the legend. Add one presidio symbol to the map near every city.

North American Deserts

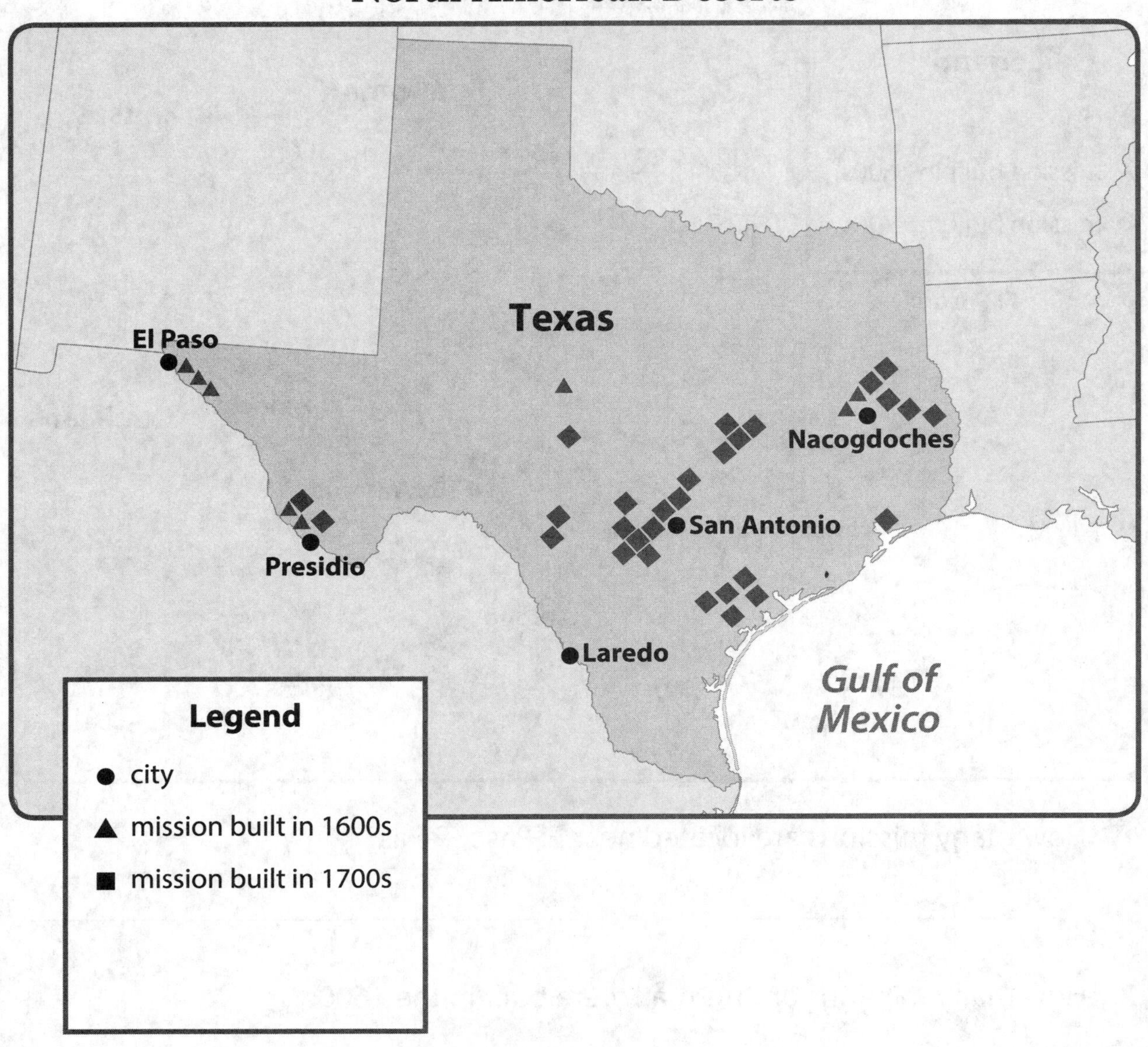

Challenge: Label the states and country that border Texas.

Name: ______________________ Date: ______________

Directions: Read the text, and study the photo. Then, answer the questions.

The Alamo

A cultural landscape is a type of building. It reflects the culture it came from. One key cultural landscape in Texas is the Alamo.

The Alamo started out as a Spanish mission. Mexico later took control of the mission. They used it as a fort. This fort helped protect Mexico from enemies.

A famous battle took place there in the early 1800s. Texas was fighting for freedom from Mexico. Texas lost this battle. Yet many soldiers chose to fight knowing they would surely lose. Their sacrifice inspired Texans. In a later battle, people were known to shout, "Remember the Alamo!"

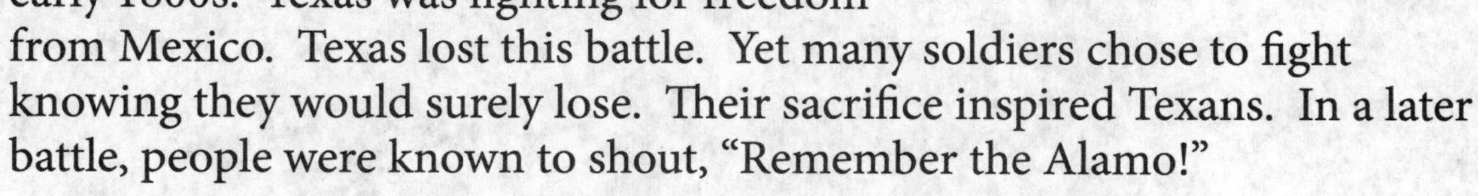

Today, the Alamo still stands. Millions visit the site each year. They learn about the history and culture of Texas. They take tours and walk through exhibits. It is a symbol of Texas history.

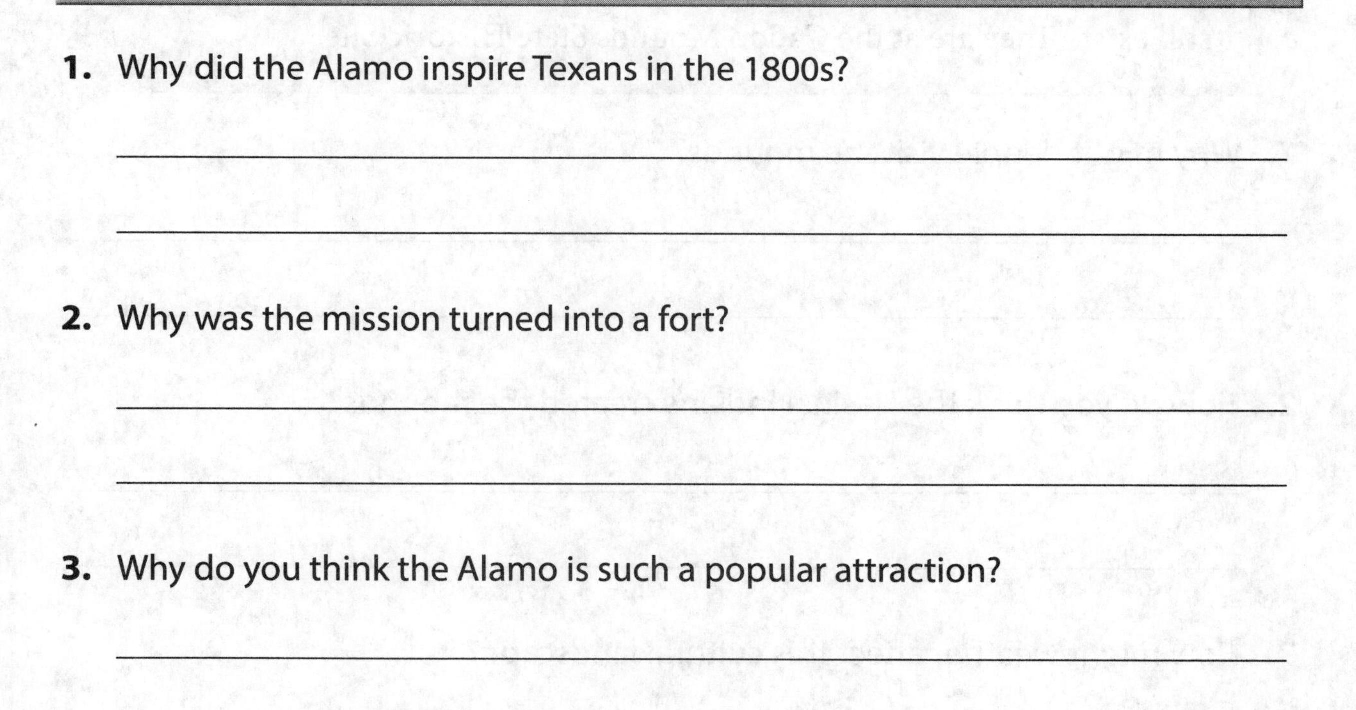

1. Why did the Alamo inspire Texans in the 1800s?

____

____

2. Why was the mission turned into a fort?

____

____

3. Why do you think the Alamo is such a popular attraction?

____

____

____

Read About It

Name: ______________________ Date: ______________

Directions: Study the photo, and read the text. Then, answer the questions.

Over 1,000 years ago, a group of American Indians lived in what is now Texas. The Hasinai Indians built mounds. They were used for ceremonies and as burial sites. Today, three of these cultural landscapes still exist. They are at the Caddo Mounds State Historic Site.

1. Why might people visit the mounds?

____

____

2. How do you think the Hasinai Indians created the mounds?

____

____

3. How might weather affect this cultural landscape?

____

____

Name: ______________________ **Date:** ______________

Directions: People bring their culture with them whenever they move to a new place. Think about some of the things you and your family do together. Write about them in the web.

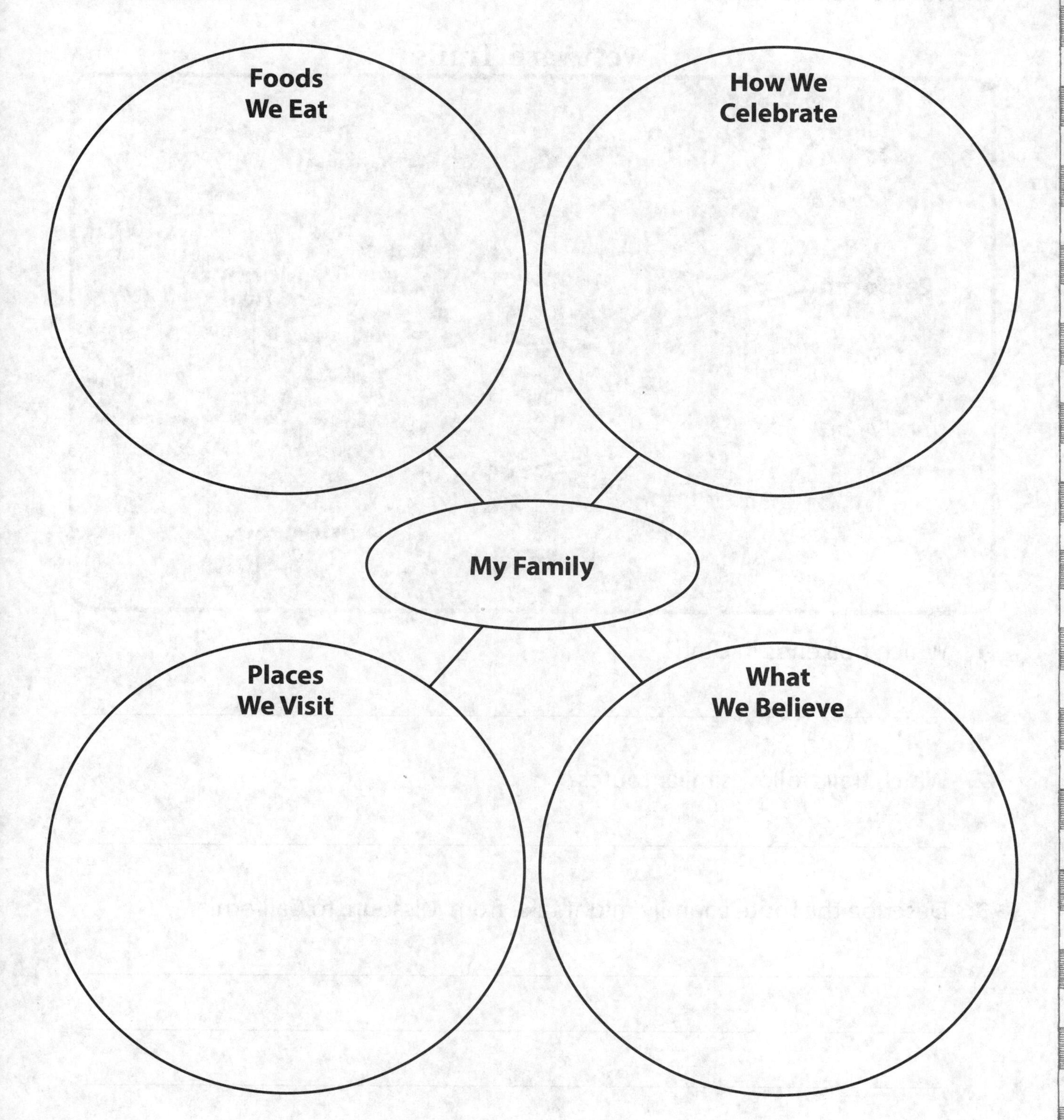

Name: ______________________ **Date:** ______________

Reading Maps

Directions: In the 1800s, many people migrated, or moved, west to start new lives. This map shows some of the routes they took west. Study the map, and answer the questions.

Westward Trails

1. Which trail ends in Utah?

____

2. Which trails follow similar routes?

____

3. Describe the route a family might take from Missouri, to California.

____

____

____

Name: ______________________________ Date: ________________

Directions: Use the clues to label the Oregon Trail, Mormon Trail, Santa Fe Trail, and California Trail.

Westward Trails

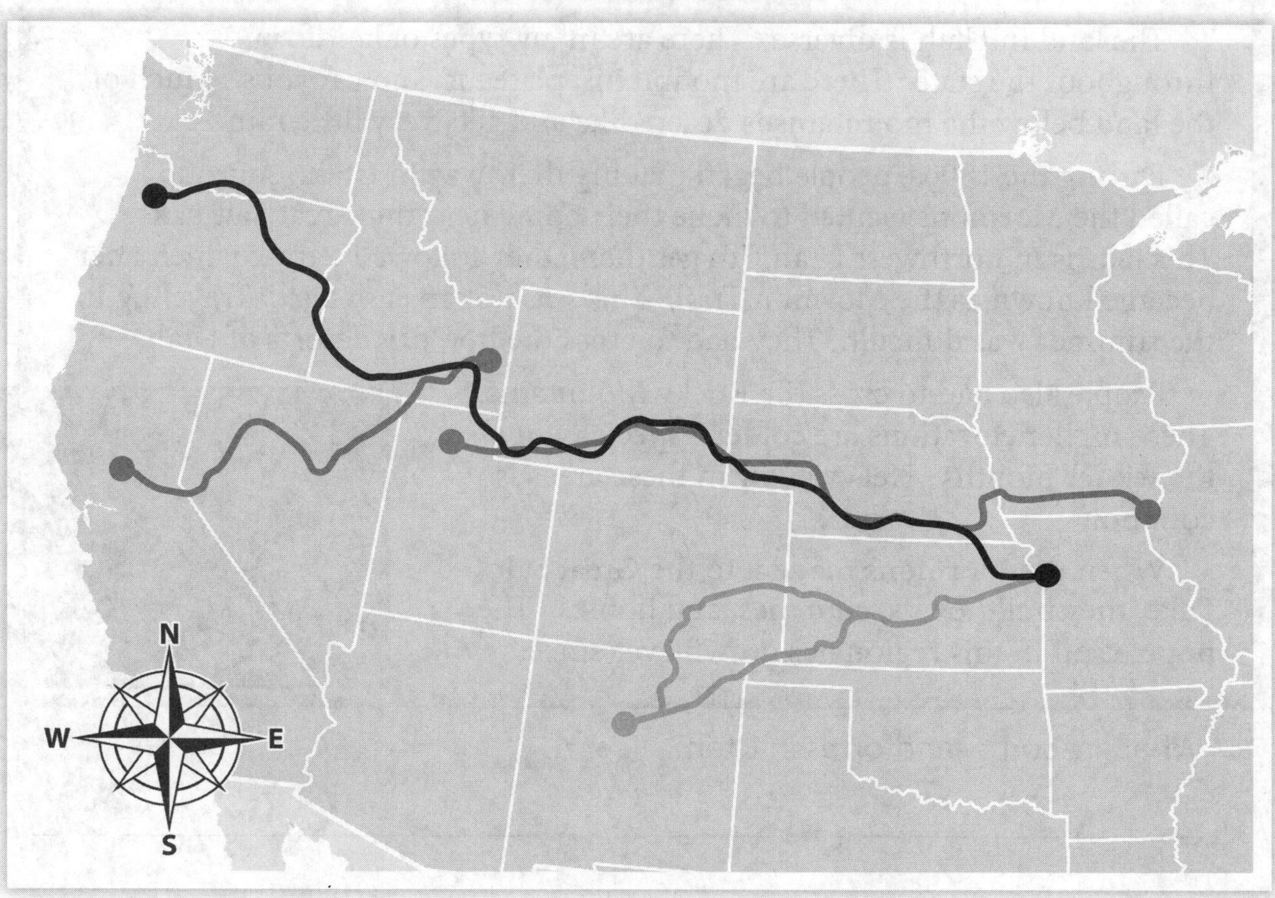

1. The Oregon Trail ends in the northwestern part of the United States.
2. The Santa Fe Trail begins in the same place as the Oregon Trail.
3. The California Trail goes through only three states.
4. The Mormon Trail closely follows the Oregon Trail in two states.

Challenge: Label as many states on the map as you can. Then, check your answers.

Name: ______________________ Date: ____________

Read About It

Directions: Read the text, and study the photo. Then, answer the questions.

Utah

The land in Utah is diverse. There are many types of landforms throughout the state. There are mountains, plateaus, and canyons. Much of the land below the mountains is desert-like and gets very little rain.

During the 1800s, people began making their way to Utah. A group called the Mormons wanted to make their home near the Great Salt Lake. This lake is in northwest Utah. To get there, they followed a trail, which later became known as the Mormon Trail. This was not an easy task. Traveling in the summer was difficult. They had to cross the hot, dry deserts of Utah.

People also had to cross the Rocky Mountains. These higher elevations are colder, especially in the winter months. Heavy snow in these areas is common.

When the Mormons made it to the Great Salt Lake, they built towns, churches, and homes. The population in this region has grown ever since.

1. What are some landforms in Utah?

____

____

2. Where did the Mormons want to start their community?

____

3. What challenges would people face crossing an area like the one in the photo?

____

____

____

Name: ______________________ Date: ______________

Directions: Railroads became important in the expansion of Utah. Railroads allowed people to start new cities and continue to move west. Study the graph, and answer the questions.

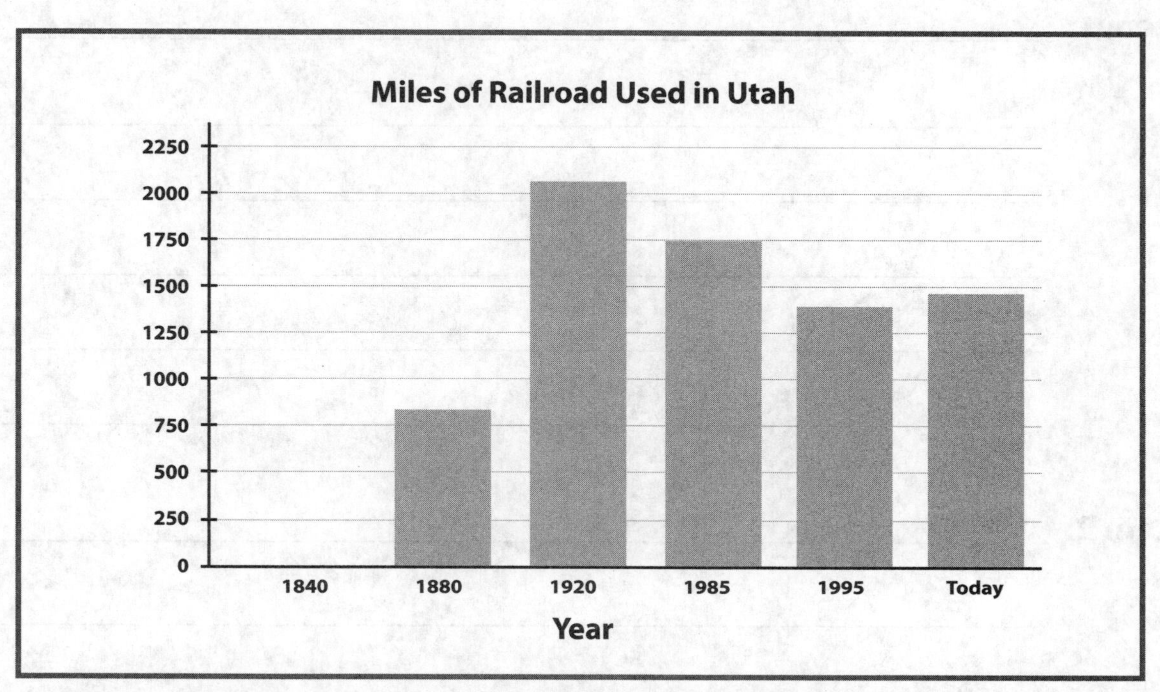

1. About how many more miles of track did Utah have at its peak compared to today?

____

2. How do you know that people did not rely on trains in 1840?

____

____

3. How do you think the invention of cars and trucks affected the railroads? What makes you think so?

____

____

____

Name: ______________________________ Date: ________________

Geography and Me

Directions: Imagine you are moving to a new area that has not been settled. Besides clothes, food, and other necessities, you are only allowed to bring three personal items on your journey. What three items would you choose? Why?

Item 1: ______________________________

____

____

____

____

____

Item 2: ______________________________

____

____

____

____

____

Item 3: ______________________________

____

____

____

____

____

Name: ______________________________ Date: ______________

Directions: This map shows the national forests in Washington and Oregon. Study the map, and answer the questions.

National Forests in Washington and Oregon

1. How many national forests are labeled in Washington?

____

2. How many national forests are labeled in Oregon?

____

3. Which national forest is farthest south?

____

4. Which national forest is in both states?

____

Creating Maps

Name: ______________________________ **Date:** ____________________

Directions: Use the chart to label the national forests in Washington.

National Forests in Washington

National Forest	Location
Umatilla	southeastern Washington
Olympic	northwestern Washington
Gifford Pinchot	southern Washington extending to central Washington
Mt. Baker-Snoqualmie	west of Okanogan and Wenatchee
Okanogan and Wenatchee	northern Washington extending to central Washington
Colville	northeastern Washington

Name: ______________________________ Date: ____________________

Directions: Read the text, and study the photo. Then, answer the questions.

Reforestation

Much of the land in Oregon and Washington is filled with forests. The wood from the trees is used to make many products. We build homes and furniture from wood. We make paper products from trees. It is important to protect our forests.

Forests are a great renewable resource. They can be replaced during our lifetime. The process of cutting trees in a forest is known as *deforestation*. Trees are cut for many reasons. People need the lumber to make things. Trees are cut to clear more space. But, it is just as important to plant new trees in their place. This is known as *reforestation*.

The process of reforestation takes time. It takes many years for trees to fully grow. States such as Washington and Oregon are doing their part. They have strict laws about reforestation. This will allow people to continue to enjoy the forests and use the resources wisely.

1. What is reforestation?

____

____

2. What are some other products made from wood not mentioned in the text?

____

____

3. Is the photo an example of reforestation or deforestation? Explain your thinking.

____

____

Think About It

Name: ______________________ Date: ______________

Directions: Wildfires can be caused by lightning strikes. These wildfires can destroy trees in a forest. This table shows how many acres have been burned in Washington due to lightning. Study the table, and answer the questions.

Year	Acres Burned
2008	1,518
2009	10,870
2010	2,397
2011	19
2012	31,425
2013	12,603
2014	265,713
2015	242,274

1. In which year was the most land burned?

____

2. Why do you think more acres were burned in some years than in than others?

____

____

3. How many more acres were burned in 2012 than in 2013?

____

4. How can people help prevent wildfires?

____

____

Name: ______________________________ **Date:** ____________________

Directions: What items in your classroom are made from wood? Draw and label them in the space below.

Reading Maps

Name: ______________________ **Date:** ______________

Directions: Study the map. Then, answer the questions.

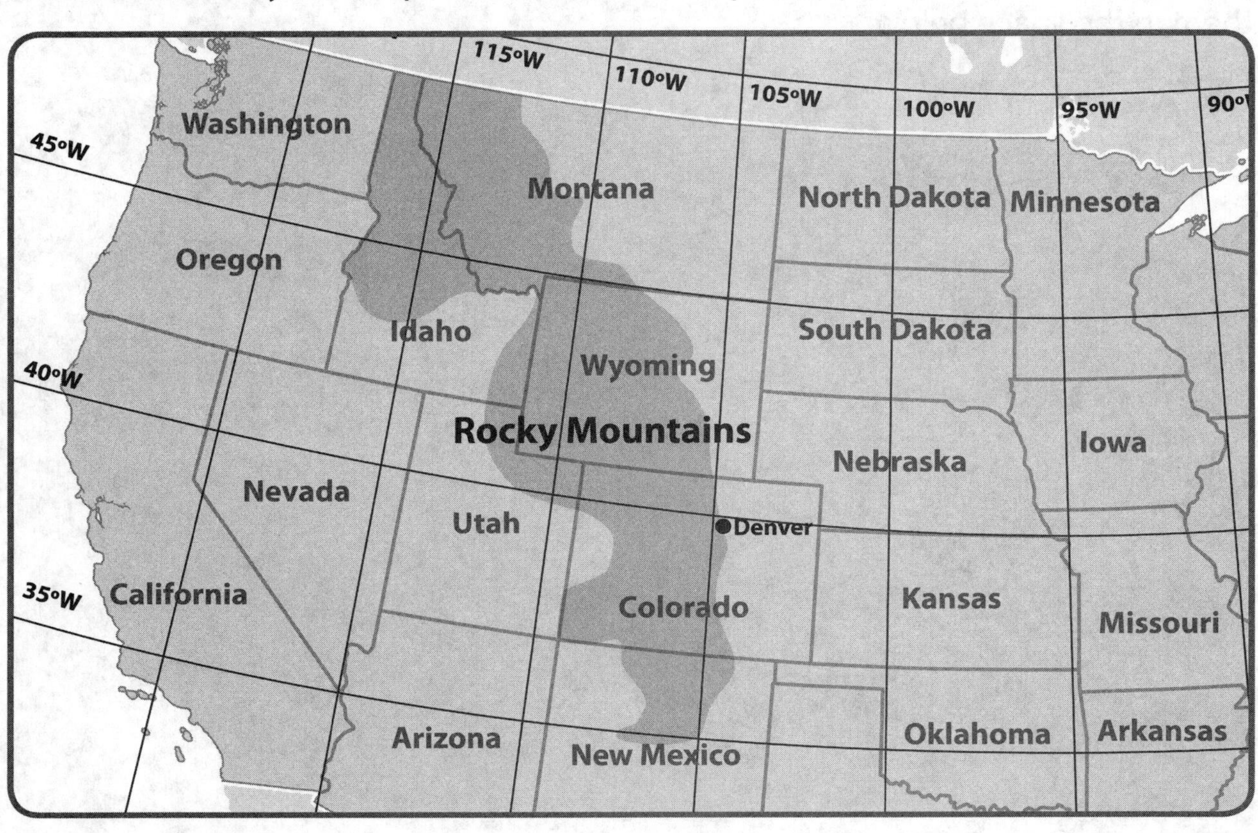

1. Which state is located at 39°N and 107°W?

____

2. What mountain range is in this state?

____

3. What other states does this mountain range pass through?

____

____

4. What are the latitude and longitude coordinates for Denver, Colorado?

____

Name: ______________________________ **Date:** ________________

Directions: Use the coordinates to correctly label each mountain on the map.

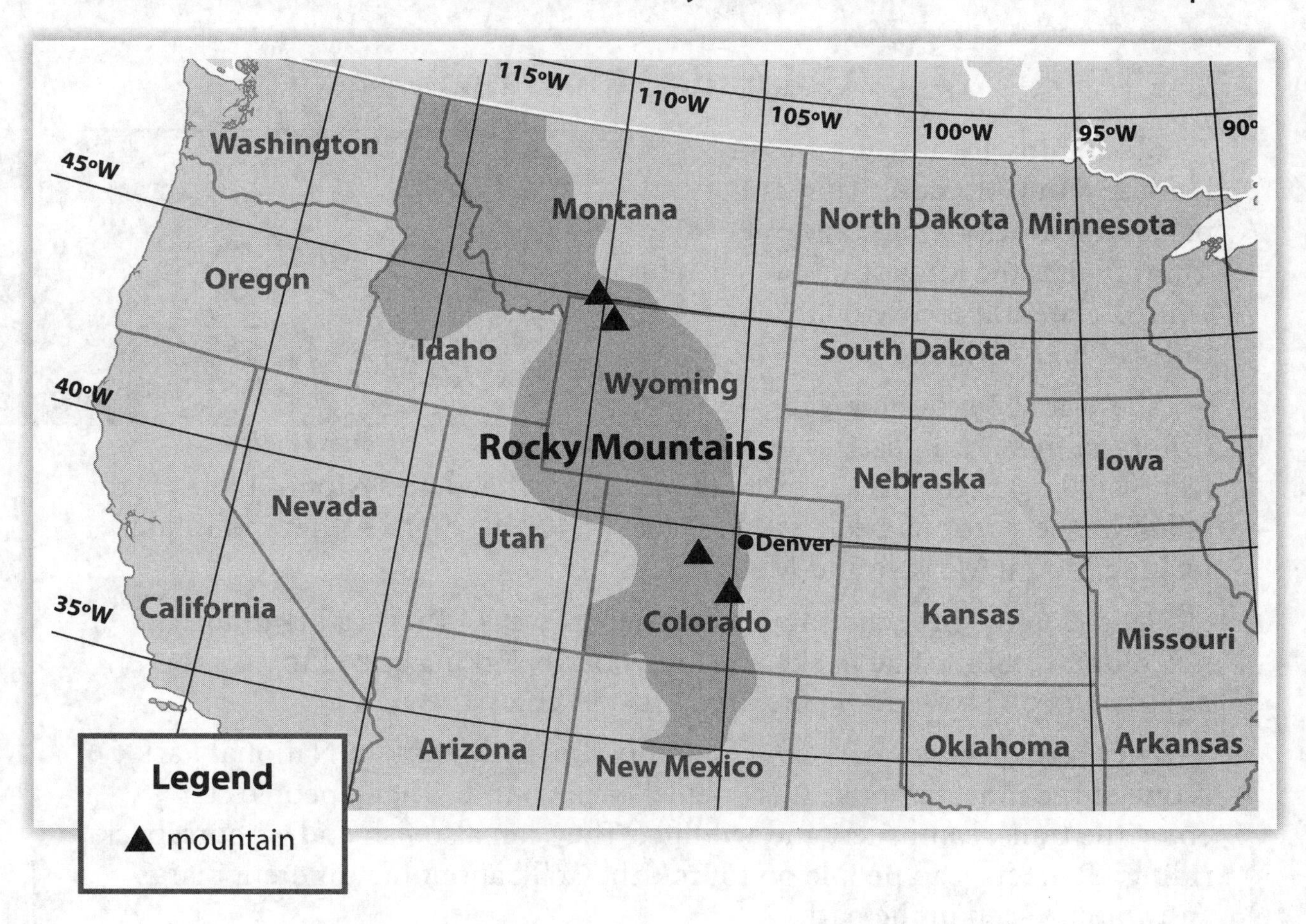

Mount Elbert: 39°N, 106°W

Francis Peak: 44°N, 109°W

Granite Peak: 45°N, 110°W

Culebra Peak: 37°N, 105°W

Challenge: Cover the names of the states with sticky notes. Write the state names on the sticky notes. Try to complete as many as you can. Then, check your answers.

Name: ____________________ Date: ____________

Read About It

Directions: Read the text, and study the photo. Then, answer the questions.

Colorado Mountains

Mount Elbert

Mountains are a major landform in Colorado. They are part of the Rocky Mountains. This range is the longest in the United States. It is over 3,000 miles (4,800 km) long.

There are 58 mountains in Colorado that are at least 14,000 feet (4,300 m) high. This is more than in any other state. Mount Elbert is the tallest in the state. Its peak is 14,443 feet (4,402 m) high. Other mountains include Mount Massive and Mount Harvard.

People enjoy these mountains in different ways. This is a big industry in Colorado. Some like to ski and snowboard. Hiking and climbing are popular, too. People even take drives just to enjoy the views.

There are plenty of parks in Colorado. Rocky Mountain National Park is one of the most famous. It is open all year round. There, people can enjoy the trails, campsites, and wildlife. They can also fish and go horseback riding. Rangers lead people on tours. They talk about the different plants and wildlife seen in the park.

1. What is one national park in Colorado?

____

2. What are some activities people can do in the Rocky Mountains?

____

____

3. At what elevation are the people in the photo?

____

Name: ______________________ Date: ______________

Directions: Sneffels Range is a group of mountains in Colorado. The elevations of some of these mountains are shown. Study the graphic, and answer the questions.

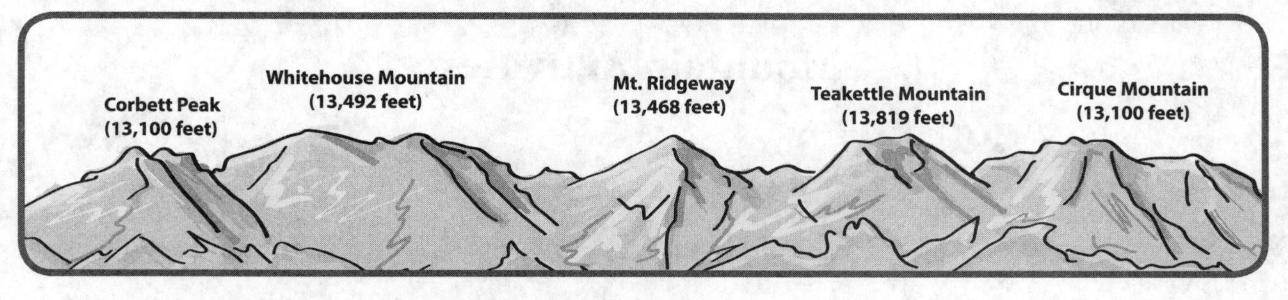

1. List the mountains from tallest to shortest.

____

____

____

2. How do the heights of these peaks compare?

____

____

____

3. Which two mountains are closest in height?

____

4. Why might someone want to visit these mountains?

____

____

____

____

Geography and Me

Name: ______________________ Date: ______________

Directions: In the box, write some activities people enjoy in the mountains. Circle one that you have tried or would like to do. Then, write about your experience or explain why you want to experience it.

Mountain Activities

-
-
-

____
____
____
____
____
____
____
____
____
____
____
____
____
____

Name: ______________________________ **Date:** ____________

Directions: Idaho has many trading partners. It sends goods to and receives products from many different countries. This map shows some of those countries. Study the map, and answer the questions.

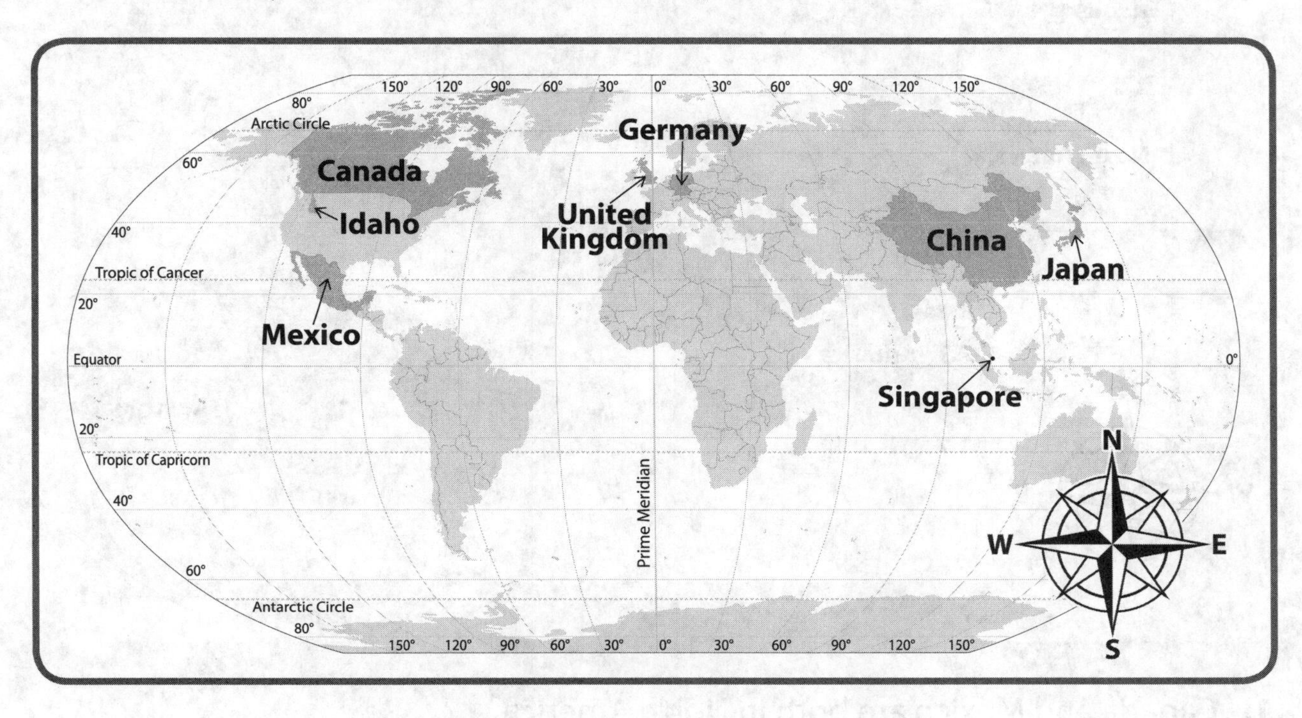

1. Which country is closest to Idaho?

____

2. Which trading partners are located east of the Prime Meridian?

____

____

____

3. What are two methods of transportation Idaho could use to send goods to the United Kingdom?

____

____

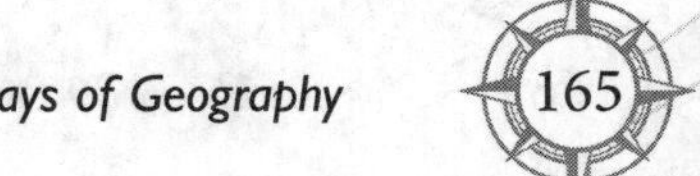

Name: ______________________________ Date: ________________

Directions: Use the clues to label countries that trade with Idaho.

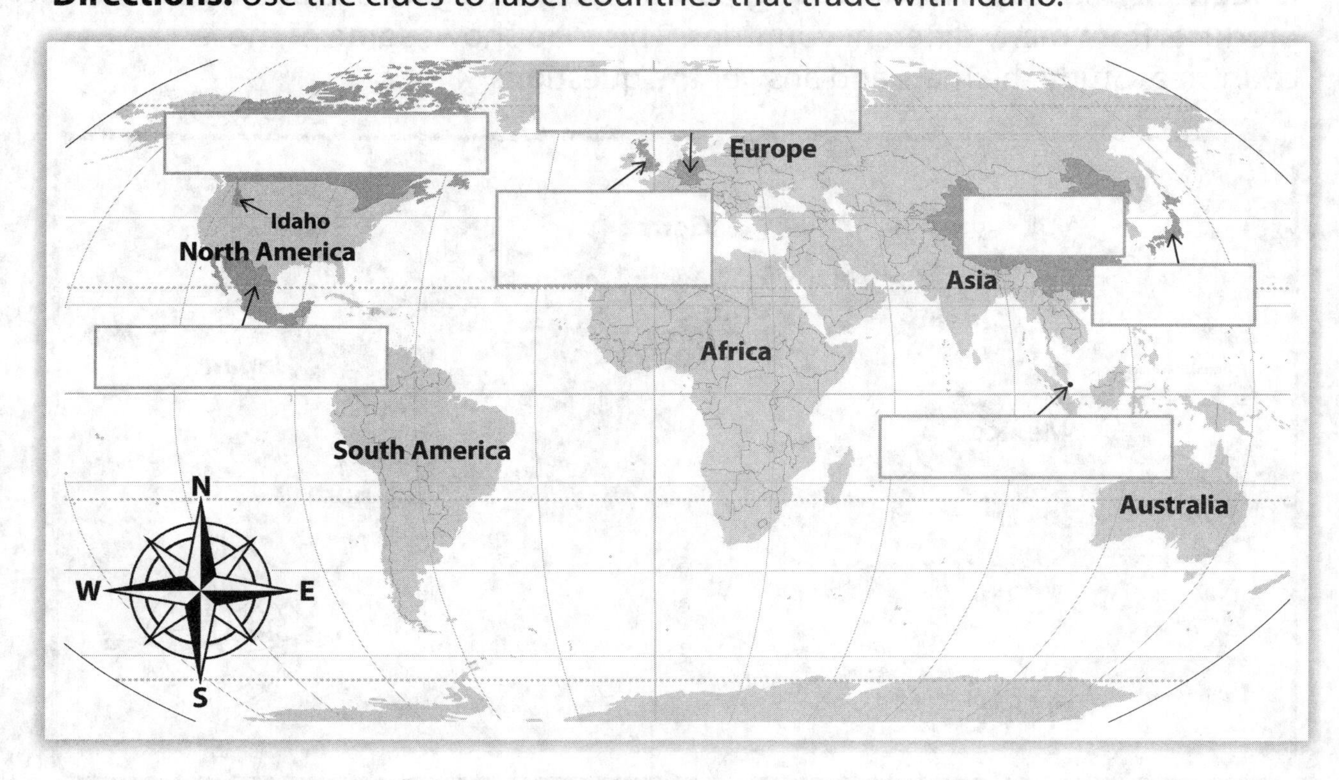

1. Canada and Mexico are both in North America.

2. Canada is north of Idaho.

3. China is a large nation in Asia.

4. Singapore is a tiny nation in Asia.

5. Japan is east of China.

6. The United Kingdom is made up of islands in Europe.

7. Germany is east of the United Kingdom.

Name: ______________________________ **Date:** ____________________

Directions: Read the text, and study the photo. Then, answer the questions.

Idaho Exports and Imports

Idaho exports, or sends, many products to other countries. The state imports, or receives, goods, too. Idaho imports and exports goods to other states as well. This brings jobs and money to the state. It also helps people get what they want and need.

Canada is a big trade partner. That means the state imports and exports goods with this country. Idaho sends wood and food products to Canada. Canada sends metals and farming products to the state.

Idaho also trades with Mexico. Idaho sends paper products to Mexico. It also exports clothes and leather. Mexico sends fruits and vegetables to the state.

Many other countries trade with Idaho. This helps the state's economy. The state brings in billions of dollars by shipping exports to other countries.

1. What goods does Idaho export to Canada?

____

2. What goods does Canada send to Idaho?

____

3. What products does Idaho send to Mexico?

____

____

4. What Idaho product is being sold in the photo?

____

Think About It

Name: ______________________________ **Date:** ______________

Directions: This graph shows the places that receive the most of Idaho's exports. Study the graph. Then, answer the questions.

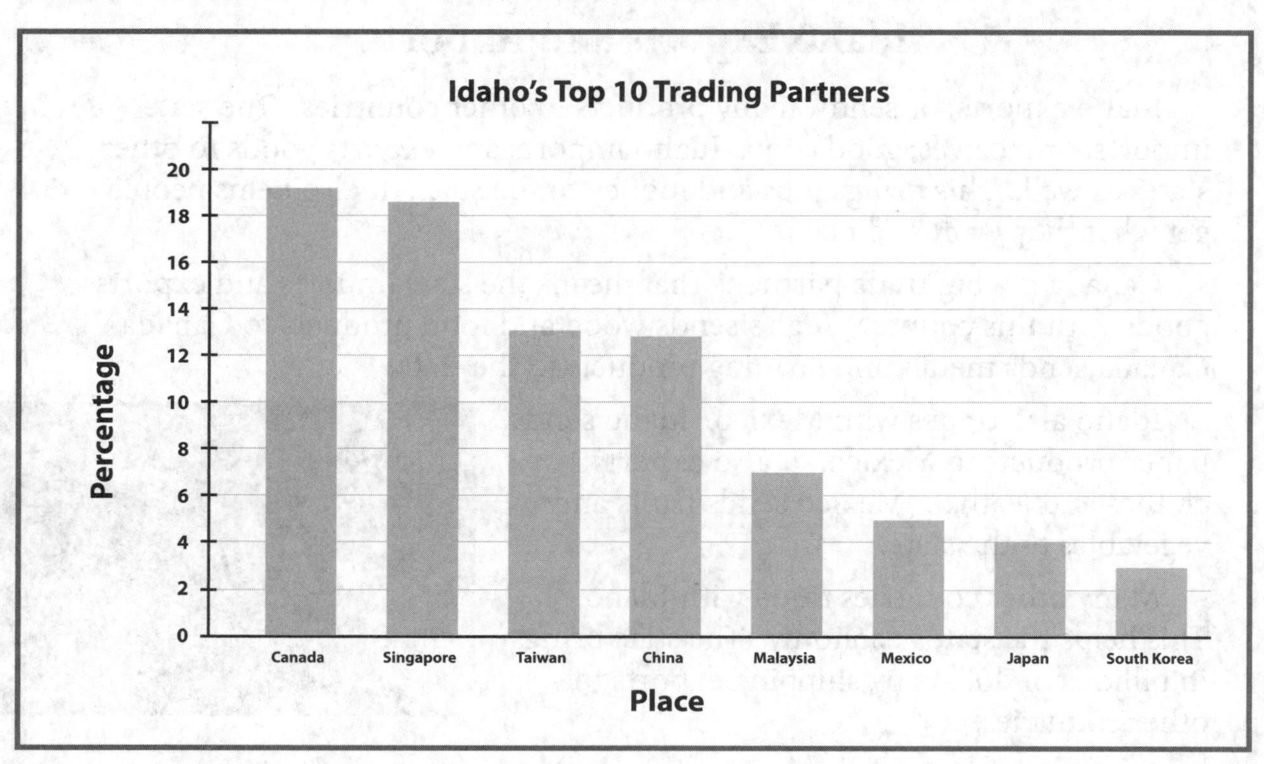

1. Which place receives the smallest amount of Idaho's exports?

____

2. Which country receives the most of Idaho's exports?

____

3. About what percentage of Idaho's exports stay in North America?

____

4. Why might the country in question 2 receive so much of Idaho's exports?

____

____

Name: ______________________________ Date: ________________

Directions: List items that you have traded with your friends or family. Draw a picture of each item you traded.

I traded ____________________ for ______________________.

I traded ____________________ for ______________________.

Name: ______________________________ Date: ________________

Directions: The Great Plains is a large area of mostly flat land. It begins in Canada and extends south to Texas. Study the map of the Great Plains. Then, answer the questions.

The Great Plains

1. How many U.S. states are part of the Great Plains?

____

2. Which states are entirely in the Great Plains?

____

____

____

3. What landform borders the Great Plains to the west?

____

Name: ______________________ Date: ______________

Directions: Label the states that make up the Great Plains. Use the word bank to help you. Then, follow the steps to complete the map.

____

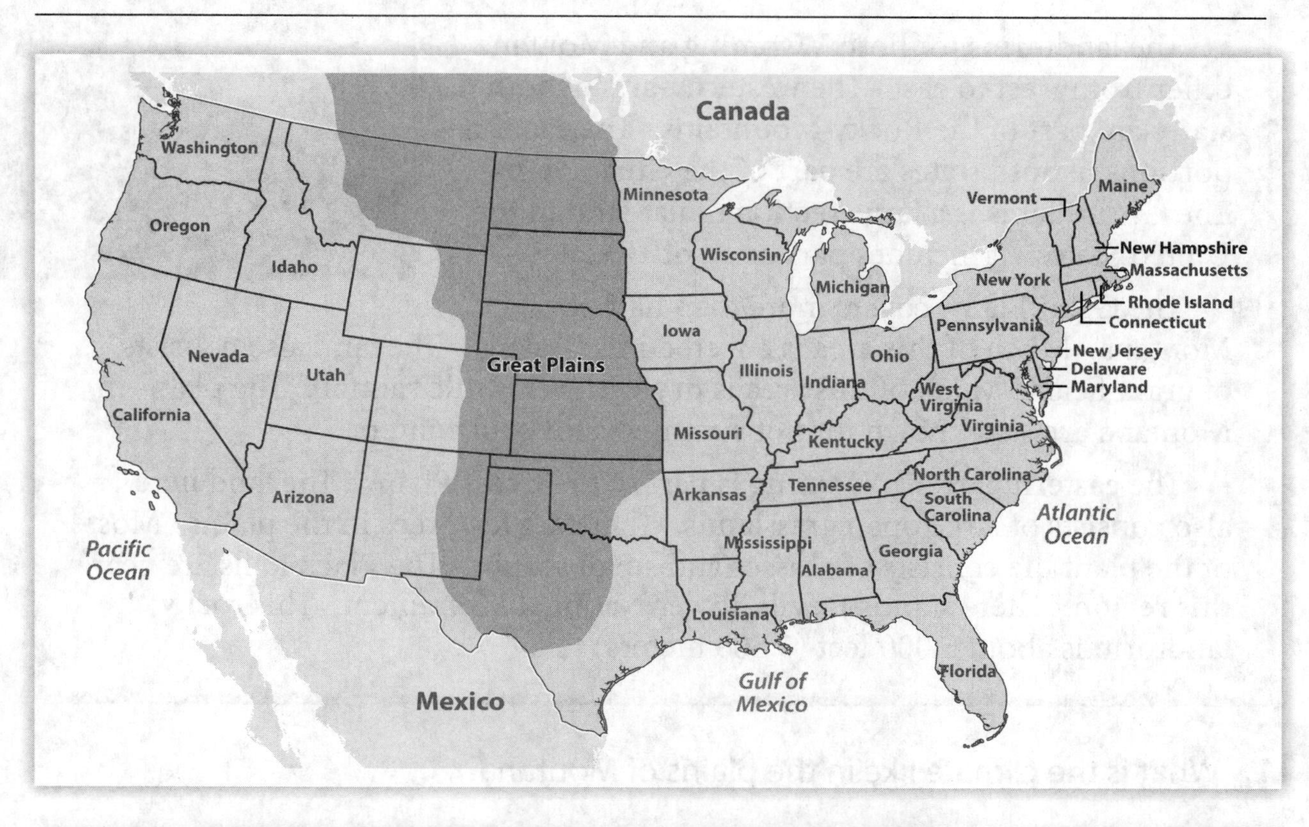

Word Bank				
North Dakota	South Dakota	Texas	Nebraska	Montana
Colorado	Kansas	Oklahoma	New Mexico	Wyoming

1. Outline the Great Plains region in red.
2. Shade the states that contain part of the Great Plains green.
3. Add a compass rose.
4. Add a title to the map.

Read About It

Name: ______________________ **Date:** ______________

Directions: Read the text, and study the photo. Then, answer the questions.

The Great Plains

The landforms for both Wyoming and Montana differ from west to east. The western parts of both states are part of the Rocky Mountains. The eastern portions of both states are part of the Great Plains. The Great Plains is a large section of flat land in the United States. It includes part or all of 10 states.

The Great Plains covers more than half of Montana. Much of this area is covered in grassland. The land also consists of grain fields. Much of this area is dry with very little rainfall. The plains in Montana are also known for hot summers and cold winters.

The eastern part of Wyoming is part of the Great Plains. The land here also consists of large open grasslands. There are few trees in the plains. Most of the plant life consists of grasses and small shrubs. The Black Hills are near this region. There stands Devils Tower National Monument. This rocky landform is about 5,000 feet (1,500 meters) tall.

1. What is the climate like in the plains of Montana?

____

____

2. What makes Devils Tower in Wyoming unique?

____

____

3. Would you rather visit the eastern or western portion of Montana and Wyoming? Why?

____

____

Name: ______________________________ Date: ________________

Directions: Temperatures in the mountains of Montana tend to be much cooler than in the plains. This chart shows the high and low temperatures of two cities in Montana. Study the chart, and answer the questions.

Month	West Yellowstone, MT	Billings, MT
Jan.	High: 24°F Low: 1°F	High: 36°F Low: 18°F
Feb.	High: 30°F Low: 4°F	High: 40°F Low: 21°F
March	High: 38°F Low: 12°F	High: 49°F Low: 27°F
April	High: 47°F Low: 21°F	High: 58°F Low: 35°F
May	High: 58°F Low: 29°F	High: 67°F Low: 44°F
June	High: 69°F Low: 36°F	High: 77°F Low: 52°F
July	High: 78°F Low: 41°F	High: 87°F Low: 59°F
Aug.	High: 77°F Low: 38°F	High: 86°F Low: 57°F

1. Which city is most likely located in the Rocky Mountains? How do you know?

____

____

2. What is the difference in low temperatures between the two cities?

____

3. What month is the warmest in both cities?

____

Geography and Me

Name: ______________________ Date: ______________

Directions: Montana and Wyoming are part of the Great Plains and the Rocky Mountains. Which region would you rather live in? Why? Draw and write to support your opinion.

____

____

____

____

____

____

____

Name: ______________________ Date: ______________

Directions: Nevada is a very dry state. This can sometimes lead to a drought. A drought is when a place does not receive enough rain. This map shows a drought in Nevada in 2016. Study the map, and answer the questions.

Drought Levels

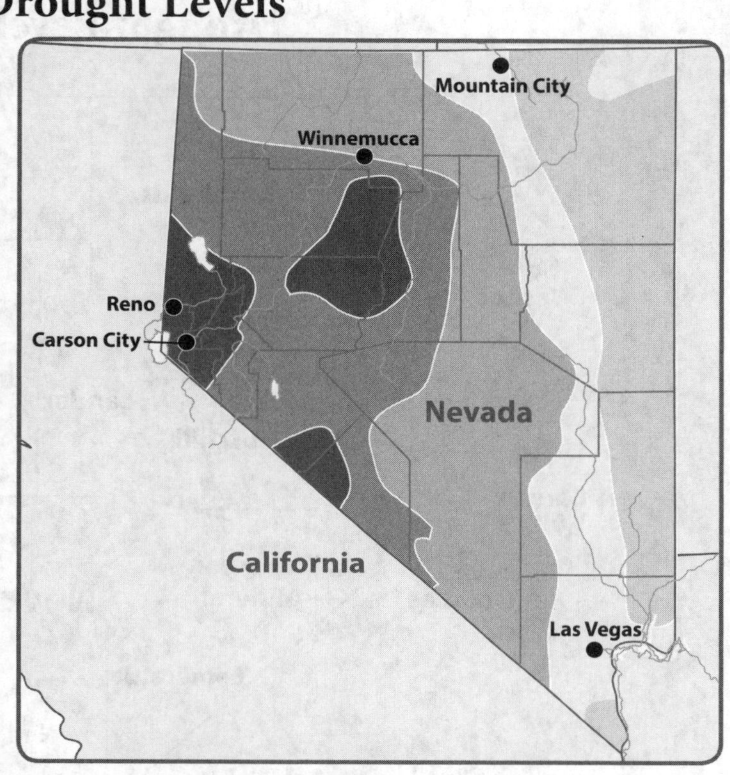

1. What is the worst level of drought? How do you know?

____

____

2. Which cities are in that level of drought?

____

3. What type of drought was Las Vegas, Nevada, experiencing in 2016?

____

4. What part of Nevada was not experiencing any type of drought in 2016?

____

Creating Maps

Name: ______________________ **Date:** ______________

Directions: The chart shows drought conditions in Nevada at the beginning of 2017. Color each county on the map according to the chart. The counties with no drought conditions will not be colored.

Drought Levels

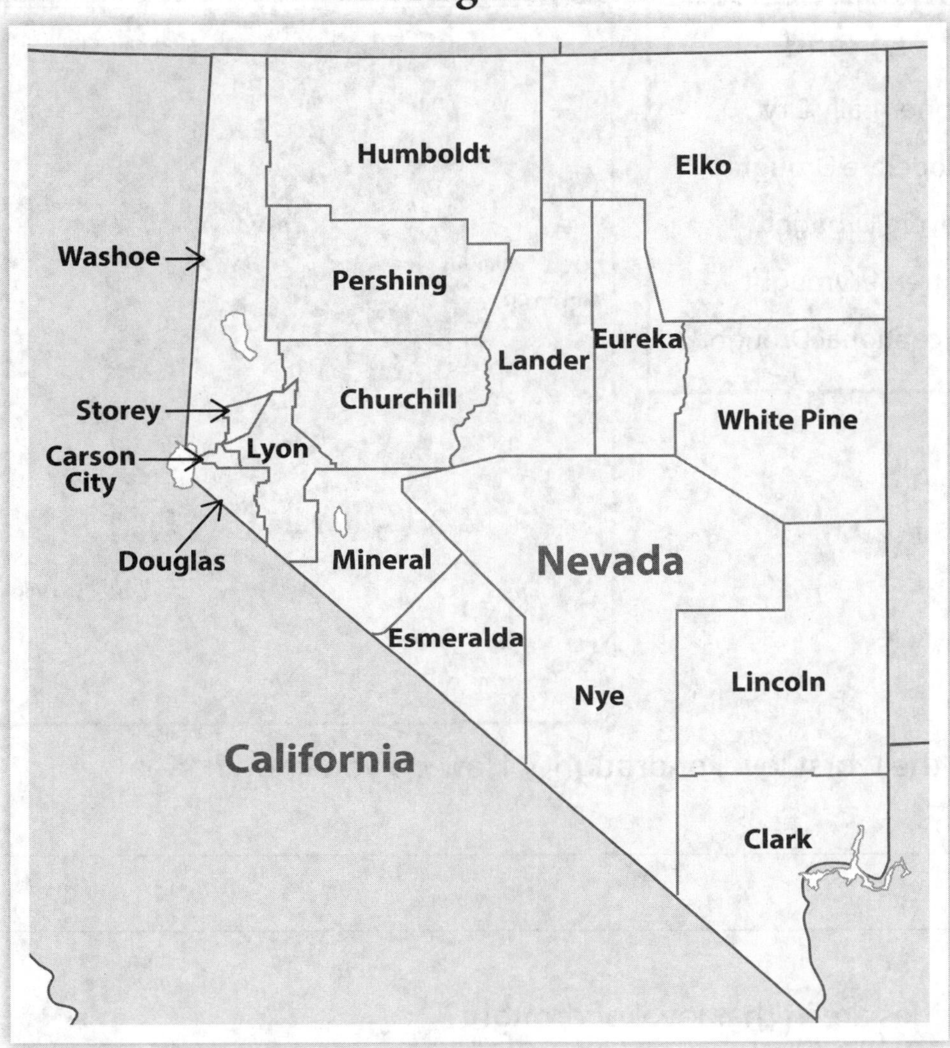

County	Drought Condition	Color
Douglas, Esmeralda, Mineral, Carson City, Lyon, Storey	severe drought	red
Washoe, Pershing, Churchill	moderate drought	orange
Humboldt, Lander, Nye, Eureka, Clark	abnormally dry	yellow

Name: ______________________ Date: ______________

Directions: Read the text, and study the photo. Then, answer the questions.

Droughts and Water Conservation

The climate in Nevada is very dry. It is not uncommon to go weeks or months with little or no rain. When there is little rainfall for a long time, a drought occurs. Droughts are common in Nevada.

Droughts affect the habitats in Nevada. Without water, plants and animals can become dehydrated. People can be affected, too. Water parks may close. Fountains may be turned off. If the drought is long enough, people may need to buy water from the store to use.

People can do a lot to conserve, or save, water. People should think twice before watering lawns during the day. It is better to water in the morning or evening when it is cooler. Another way is to use less water when washing cars. People can take shorter showers. They should only run dishwashers and washing machines when they are full. When everyone cuts back, it can make a difference.

1. What are two effects of a drought?

____

____

2. What are two ways to conserve water?

____

____

3. How can you tell that the area in the photo is experiencing a drought?

____

____

Think About It

Name: ______________________ **Date:** ______________

Directions: This graph shows the average precipitation in Eureka, Nevada. Study the graph. Then, answer the questions.

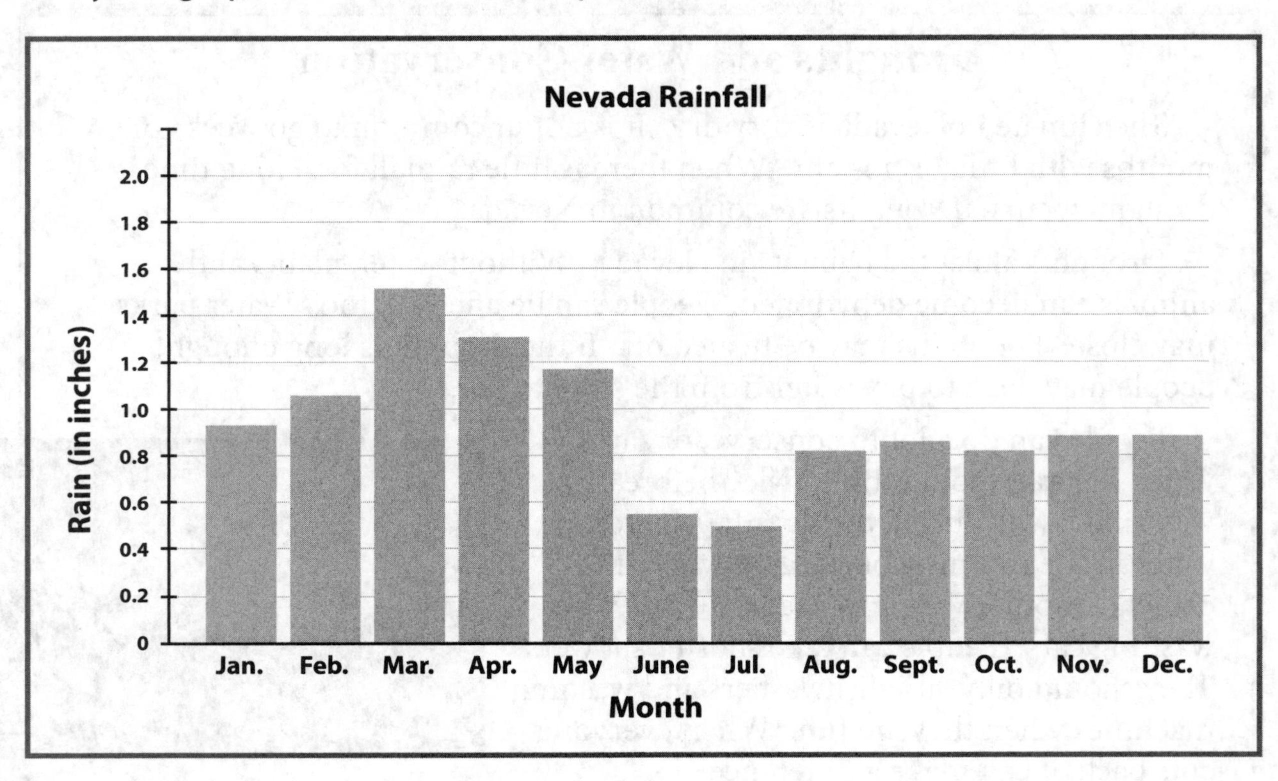

1. How many months of the year does Eureka receive at least one inch of rain?

____

2. During which season does Eureka receive the most rain? Explain your reasoning.

____

____

____

3. During which two months is Eureka most likely to experience a drought? Explain your reasoning.

____

____

Name: ______________________________ Date: ______________

Directions: What are some ways you can conserve water where you live? Make a list of indoor and outdoor water-saving tips.

Indoor	**Outdoor**

Reading Maps

Name: ______________________ Date: ______________

Directions: Earthquakes often occur along fault lines. Study the map of fault lines in California. Then, answer the questions.

California Fault Lines

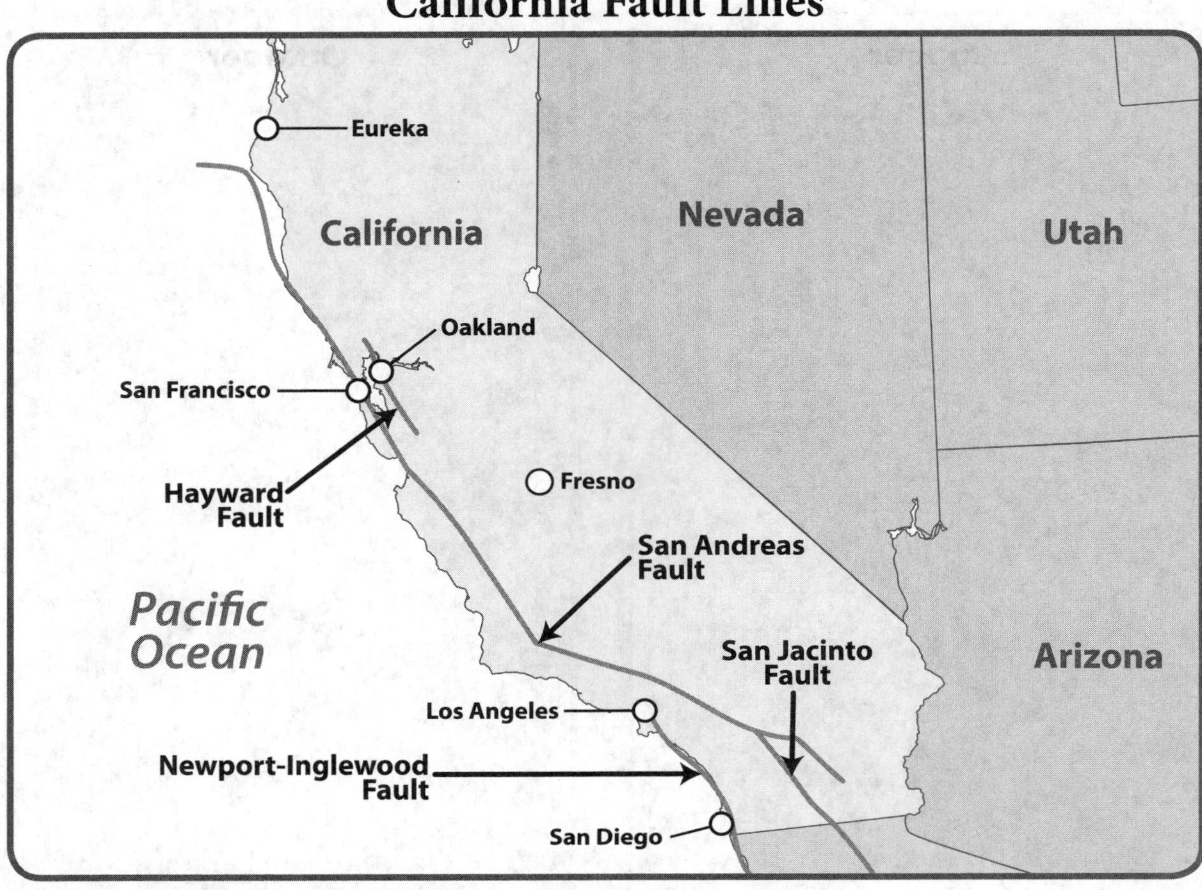

1. How many fault lines are shown on the map?

____

2. What city is on the Hayward Fault Line?

____

3. What is the longest fault line in California?

____

4. Which fault line extends between Los Angeles and San Diego?

____

Name: ________________________ Date: ______________

Directions: Use the clues to label the cities on the map.

California Fault Lines

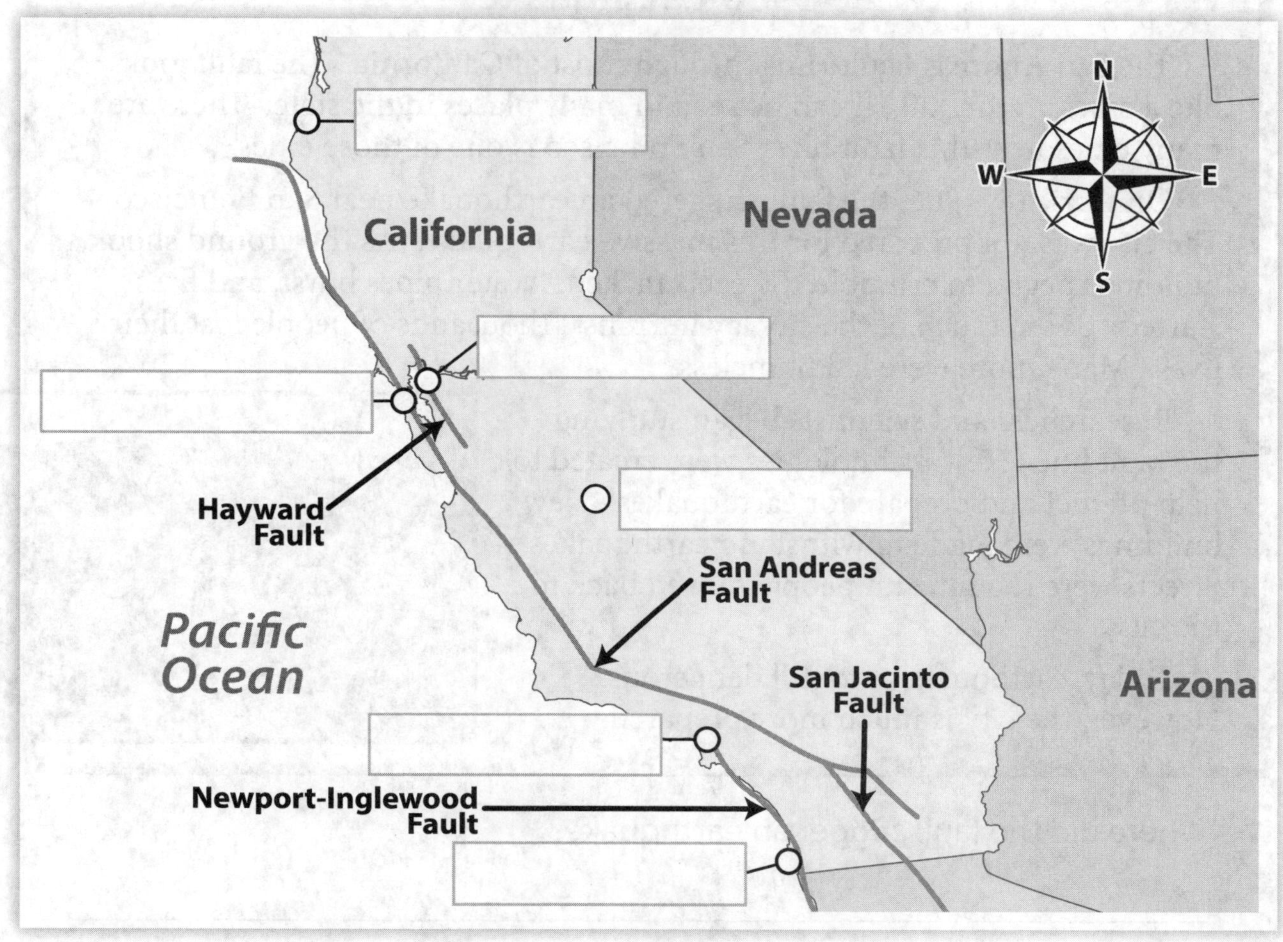

1. Eureka is in northwestern California.
2. Fresno is in central California.
3. Los Angeles is just south of the San Andreas Fault.
4. Oakland is on the Hayward Fault.
5. San Diego is along the southern part of the Newport-Inglewood Fault.
6. San Francisco is west of Oakland.

Read About It

Name: ______________________ **Date:** ______________

Directions: Read the text, and study the photo. Then, answer the questions.

1906 Earthquake

The San Andreas Fault runs through most of California. The fault looks like a crack in the soil. It can be seen in many places in the state. There are many cities near this fault line. San Francisco is one of those cities.

On April 18, 1906, the fault triggered an earthquake near San Francisco. The city was not prepared for this massive earthquake. As the ground shook, buildings began to crumble. Streets cracked. Water pipes burst, and fires started. Soon, much of the city lay in ruins. Thousands of people lost their lives. Many more were left homeless.

Researchers and scientists began studying the fault line. New technologies were created to help predict and prepare for earthquakes. New buildings were made to withstand earthquakes. Streets were rebuilt, and people moved back to the city.

Today, earthquakes are still dangerous. However, the city is much more prepared.

1. Where did the fault trigger an earthquake?

____

2. Name two effects of the 1906 earthquake.

____

____

3. How is the city better prepared today?

____

____

4. Circle the fault line in the above photo.

Name: ______________________ Date: ______________

Directions: Below is a list of supplies needed in case of an emergency, such as an earthquake. Study the list, and answer the questions.

Basic Disaster Supplies Kit

- ☑ water (three-day supply)
- ☑ food (three-day supply)
- ☑ radio
- ☑ flashlight
- ☑ first aid kit
- ☑ whistle to signal for help
- ☑ moist towelettes
- ☑ tools
- ☑ can opener
- ☑ local maps

1. What supply do you think is the most important? Explain why you think so.

____

____

____

2. Why do you think a three-day supply of food and water is suggested?

____

____

____

3. What do you think should be included in the first aid kit?

____

____

____

Name: ______________________ Date: ______________

Geography and Me

Directions: Keeping yourself safe is important during a natural disaster. Make a list of safety tips for you and your family during and after a natural disaster.

Name: ____________________ Date: ____________

Directions: There are several volcanoes in Hawai'i. Study the diagram of a shield volcano. Then, answer the questions.

Parts of a Volcano

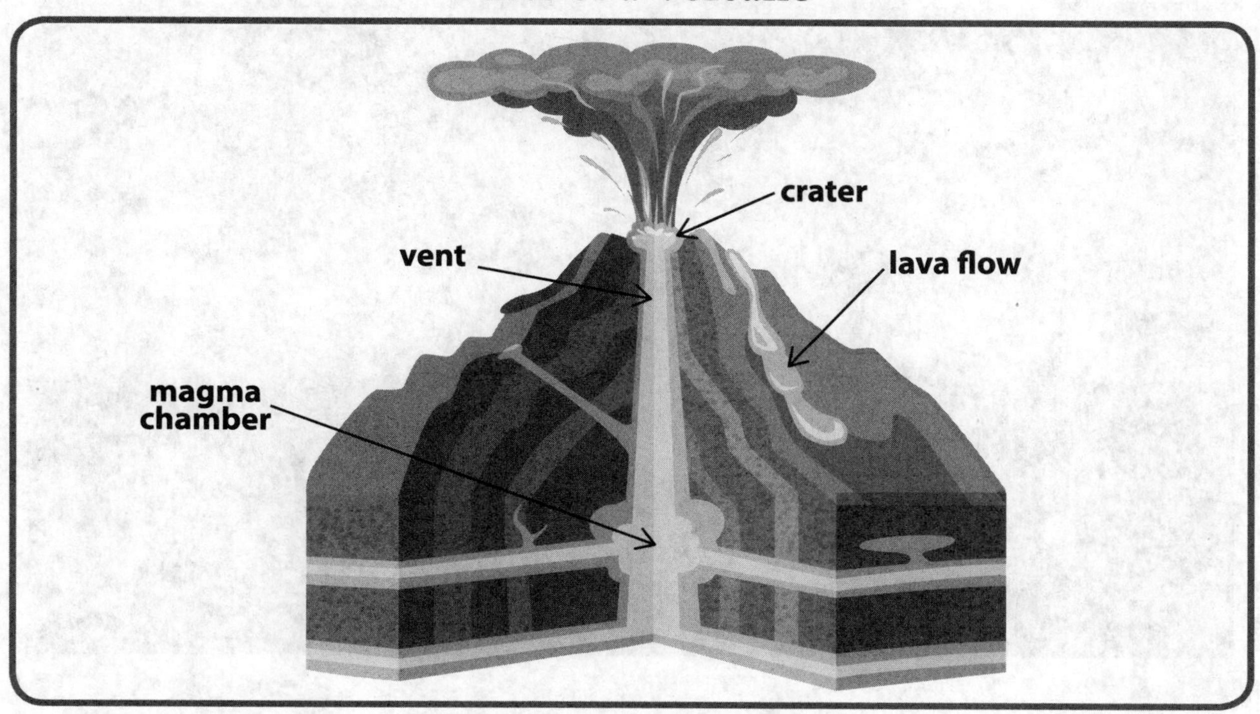

1. Where is the magma chamber located?

____

2. Where does magma exit the volcano?

____

3. Describe the shape of this volcano.

____

____

4. Describe a lava flow.

____

____

Name: ______________________ Date: ______________

Directions: This is a photo of an active volcano. Label the crater and lava. Then, describe what you might see when visiting a volcano.

____

____

____

____

____

____

____

____

____

____

____

Name: ______________________________ **Date:** ______________

Directions: Read the text, and study the photo. Then, answer the questions.

Volcanoes

The Hawai'ian Islands were formed by volcanoes. As volcanoes erupted in the ocean, they slowly got taller. Finally, they were tall enough to rise above the ocean.

Volcanoes erupt when molten, or liquid, rock under the ground is under pressure. As the pressure builds, the molten rock, or *magma*, pushes its way up through a crack in Earth's crust. Once the magma hits the surface, it is called *lava*. The lava cools and becomes solid rock again.

Most of the volcanoes in Hawai'i are shield volcanoes. These are low and wide like a shield. Lava often slowly oozes from the crater. The lava flow can last for years.

Some volcanoes in Hawai'i are active. This means they are erupting. Many people travel to the islands to see them.

1. What might make a volcano dangerous?

____

____

2. Why is it called a shield volcano?

____

____

3. How can you tell that the lava flow in the photo is recent?

____

____

Think About It

Name: ______________________ Date: ____________

Directions: Study the map of Hawai'i. Then answer the questions.

Kaua'i
Ni'ihau
O'ahu
Moloka'i
Maui
Lana'i
Haleakala
Kaho'olawe
Pacific Ocean
Hawai'i
Haulalai
Mauna Loa
Kilauea
Lo'ihi

Legend
▲ volcano

1. Where are most of the volcanoes located?

____

2. Tourists visit Kilauea's active lava flow daily. How do you think this is possible?

____

____

3. Why might someone want to visit a volcano?

____

____

Name: ______________________________ **Date:** ____________________

Directions: People visit Volcanoes National Park in Hawai'i every day. They enjoy hiking near and on the volcanoes. Draw and label some items you would bring if you were going hiking with your family at this national park.

Name: ______________________________ **Date:** ______________

Directions: Much of the land in Alaska is tundra. This land has freezing temperatures for most of the year. Study the map of Alaska. Then, answer the questions.

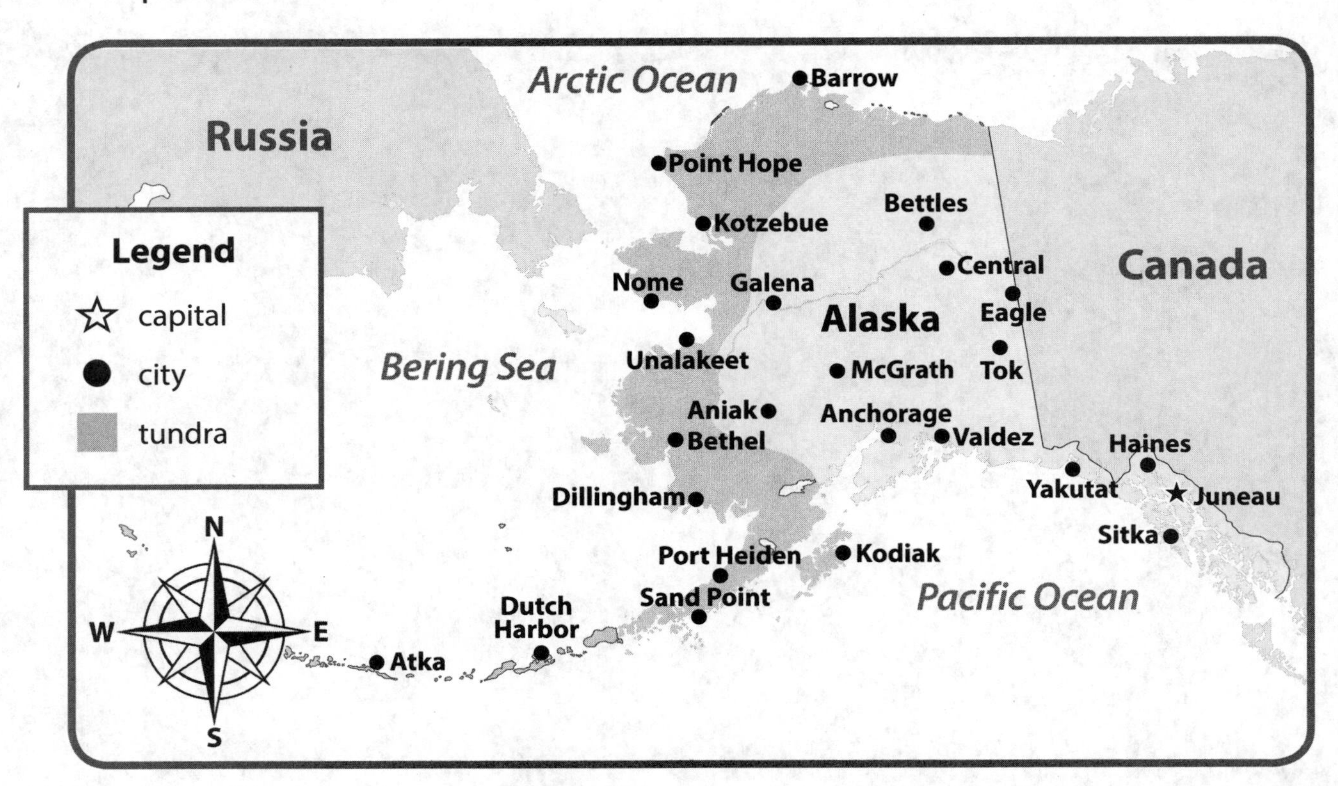

1. What is the capital of Alaska?

____

2. What oceans and seas border Alaska?

____

3. What country is west of Alaska?

____

4. Describe the location of the tundra in Alaska.

____

____

Name: ______________________ **Date:** ______________

Directions: Follow the steps to complete the map.

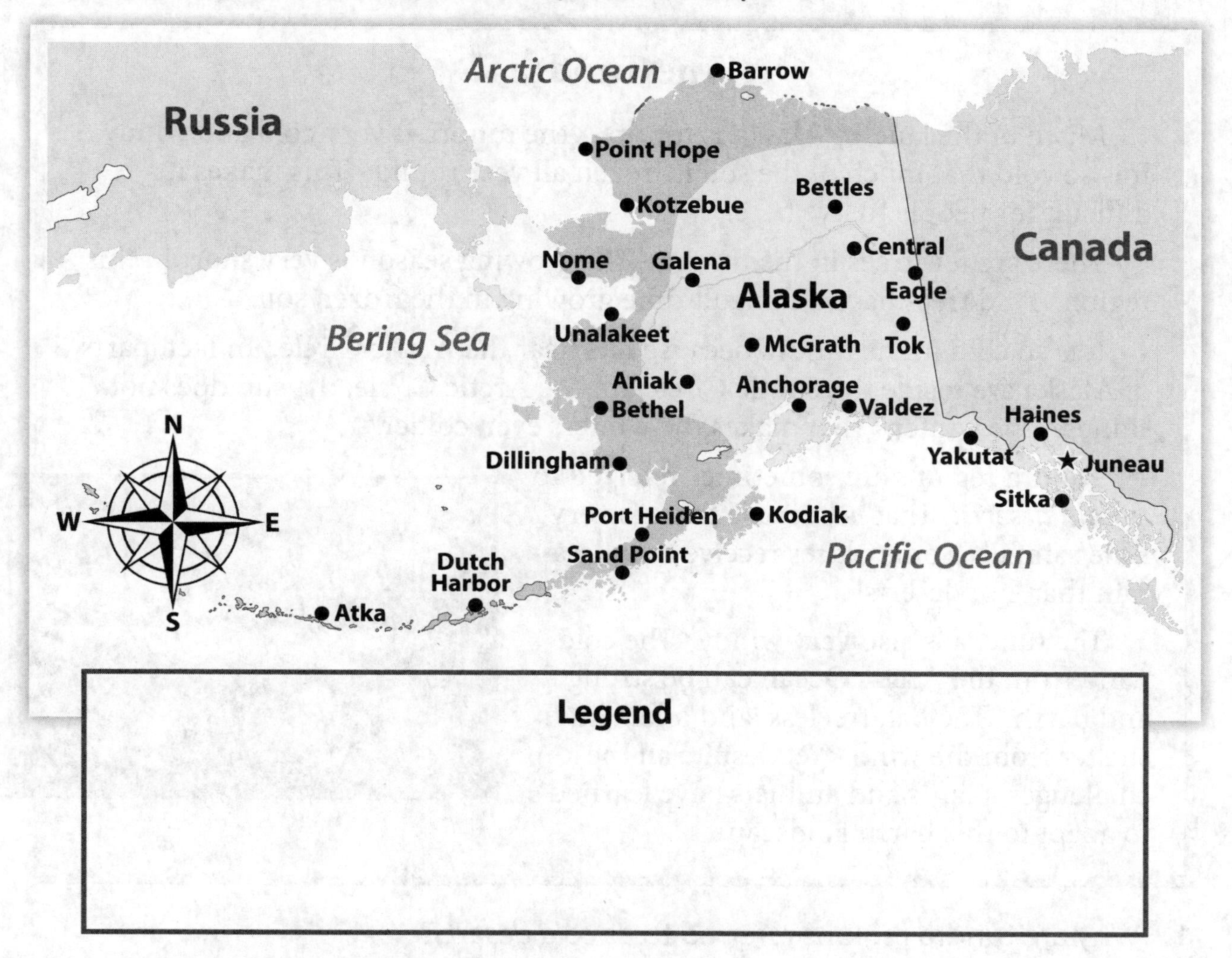

Legend

1. Color the tundra purple.

2. Color Russia green.

3. Color Canada orange.

4. Color the oceans blue.

5. Create a legend for the map. Include the colors you used and what they represent.

Name: ______________________ Date: ______________

Read About It

Directions: Read the text, and study the photo. Then, answer the questions.

The Tundra

Much of the land in Alaska is tundra. The tundra is very cold and windy. It is so cold that much of the soil is frozen all year round. This makes it difficult for people to live here.

There are few trees in the tundra. The growing season is very short in this region. And trees have a difficult time growing in the frozen soil.

It is so cold in the tundra because it is near the Arctic Circle. In fact, parts of Alaska are inside the Arctic Circle. In the Arctic Circle, the sun does not shine in the winter. That makes the winters even colder.

Tundra regions are sometimes referred to as *cold deserts*. That is because there is very little rainfall. Some places receive even less rain than hot deserts!

The tundra is also very windy. The cold wind from the Arctic Ocean can be strong and harsh. The flat, treeless land offers little shelter from the wind. Yet despite all these challenges, plants and animals have learned to adapt to this harsh landscape.

1. Why are tundra regions referred to as *cold deserts*?

____

____

2. Why do trees have a hard time growing in the tundra?

____

____

3. Where does much of the wind in the Alaskan tundra come from?

____

Name: ______________________ **Date:** ______________

Directions: Study the photo of the tundra in Alaska. Then, answer the questions.

1. How can you tell this is part of the tundra?

____

____

2. Some American Indian and Alaska native tribes live in the tundra. How do you think they survive here?

____

____

3. Do you think there are farms in a tundra? Explain your answer.

____

____

Name: ______________________ Date: ______________

Directions: Write where you live in the Venn diagram. Then, compare and contrast the climate of a tundra with the climate in your area.

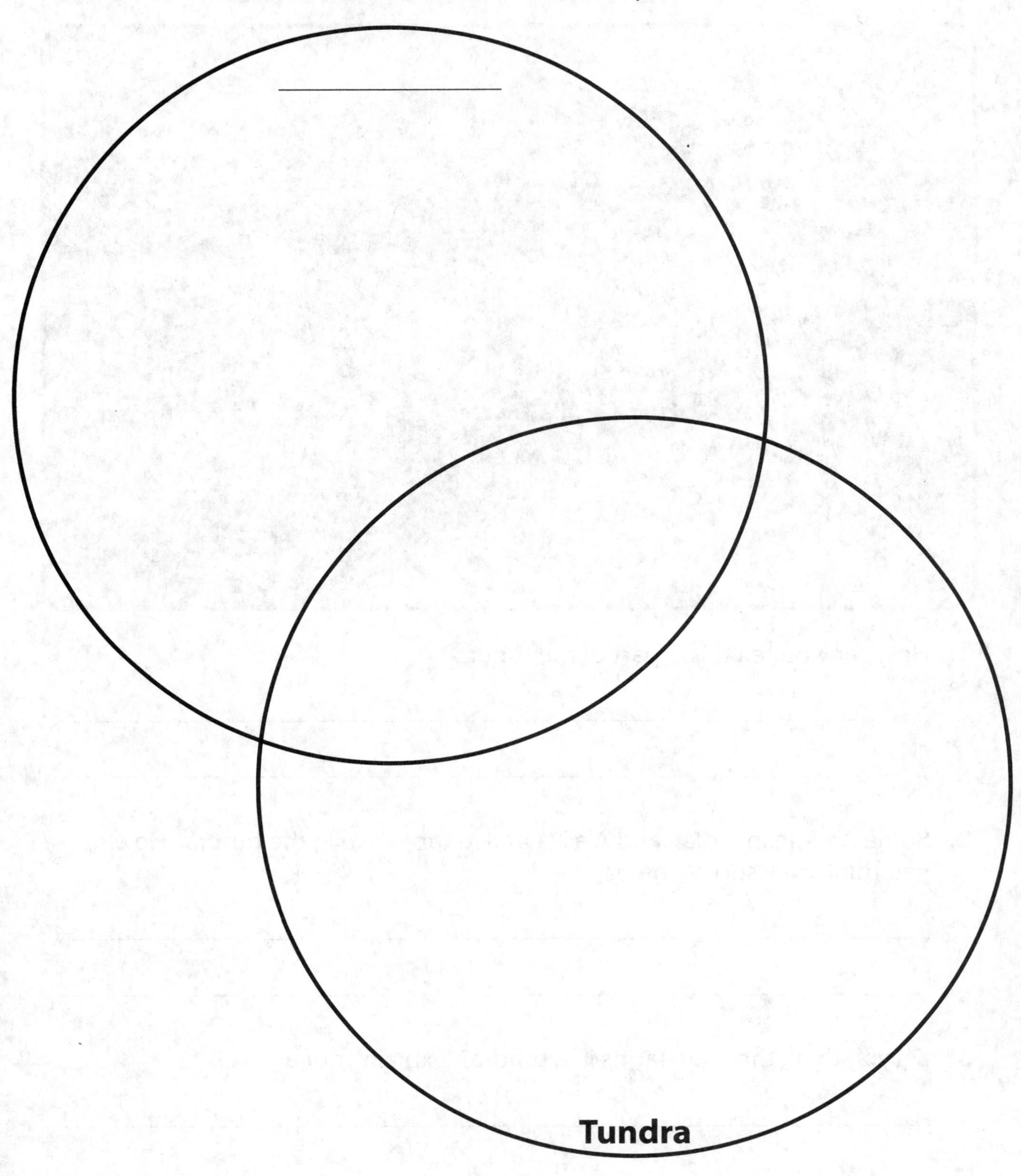

ANSWER KEY

There are many open-ended pages and writing prompts in this book. For those activities, the answers will vary. Answers are only given in this answer key if they are specific.

Week 1 Day 1 (page 15)

1. north
2. east
3. southeast
4. southwest
5. northeast
6. northwest
7. south

Week 1 Day 2 (page 16)

1. Illinois
2. California
3. 45°N, 100°W

Week 1 Day 3 (page 17)

1. 400 miles
2. 600 miles
3. 800 miles
4. 1,000 miles

Week 1 Day 4 (page 18)

1. Canada
2. Arctic Ocean and Pacific Ocean
3. Los Angeles

Week 1 Day 5 (page 19)

1. Rocky Mountains and Appalachian Mountains
2. Answers may include that it is in Canada or that it is very far north.
3. Answers should indicate that it is in the southwestern United States.

Week 2 Day 1 (page 20)

1. Hawai'i, Florida
2. 56°; Virginia's average temperature is in the 50s.
3. Answers may include that it is colder farther north or that temperatures are similar in nearby states.

Week 2 Day 2 (page 21)

1. highways 405 and 605
2. highway 10, 60, 605
3. B3

Week 2 Day 3 (page 22)

1. Rio Grande
2. Gulf of Mexico
3. Colorado, Kansas, Oklahoma, and Arkansas
4. Colorado River

Week 2 Day 4 (page 23)

1. 11 states
2. Oregon, California
3. Indian Territory

Week 2 Day 5 (page 24)

1. 9 states
2. South
3. Ohio
4. South

Week 3 Day 1 (page 25)

1. Bar Harbor, Portland
2. salmon, crab
3. Answers may include that they could drive to a coastal city or that seafood can be delivered.

Week 3 Day 2 (page 26)

Maps should have the following drawings from the legend: Boston: clams, cod, and bass; New Bedford: mussels; Gloucester: clams and crab; Salem: cod and crab; Plymouth: bass.

Week 3 Day 3 (page 27)

1. A coastal state is a state that borders an ocean.
2. Answers should include two of the following: fisher, store clerk, delivery person, chef, or server.
3. Answers may include that lobsters are long or that fishers can catch more lobster in the traps.

Week 3 Day 4 (page 28)

1. Answers may include that the fish stays fresher, there is less travel time, or there is a larger variety of seafood.
2. Answers may include that fewer people are needed to bring the seafood to the market, there is an abundance of seafood near the shore, or there are more markets selling seafood near the shore.
3. Answers may include cleaning the fish, advertising, setting prices, or displaying the fish.

Week 4 Day 1 (page 30)

1. east
2. D3
3. Answers may include that it is safer to swim in B1 because it is closer to shore.

ANSWER KEY *(cont.)*

Week 4 Day 3 (page 32)

1. Answers may include that humans have taken over their habitats, and litter and pollution can harm plants and animals.
2. Answers may include the buildings in the background.
3. Answers may include playing sports, walking, running, and playing in the sand.

Week 4 Day 4 (page 33)

1. 30 million
2. More; answers should explain that bars are higher over time.
3. 10 years

Week 5 Day 1 (page 35)

1. electronics
2. Example: *Head south on Main Street. Turn right, and go west on 2nd Street.*
3. Answers may include that it takes up less space, workers can stay in one area, and it keeps trash and recyclables in one place.

Week 5 Day 3 (page 37)

1. light and noise pollution
2. air pollution, noise pollution, or water pollution
3. Answers may include picking up trash and walking instead of driving.

Week 5 Day 4 (page 38)

1. 16,222 complaints
2. 4,106 complaints
3. Example: *No, cars and trucks are always on the streets, and there is a lot of construction in a city.*

Week 6 Day 1 (page 40)

1. Canaan
2. Winter; answers should include that winter is the coldest season and there are subzero temperatures.
3. Answers should reference the temperatures shown on the map.

Week 6 Day 3 (page 42)

1. Winter; answers may include that most plants have lost their leaves, and many animals are hibernating.
2. Answers may include that the temperatures drop and leaves change color.
3. Answers may include that people stay indoors more when it is cold, they need to wear different clothes for different seasons, or they may do different activities.

Week 6 Day 4 (page 43)

1. March
2. Answers should indicate one of the summer months.
3. Answers may include that the graph will not change because the seasons follow a yearly cycle.

Week 7 Day 1 (page 45)

1. Ridge and Valley; a star
2. Allegheny Plateau
3. Atlantic Coastal Plain, Piedmont, Ridge and Valley, and Allegheny Plateau

Week 7 Day 3 (page 47)

1. Miners used hand tools and carried the coal out.
2. Answers may include that they needed money or that coal was an important resource.
3. Answers may include that they are dirty or they are children.

Week 7 Day 4 (page 48)

1. Answers may include hard hats or safety vests.
2. Answers may include that it prevents things from falling onto their heads.
3. Answers may include to give them ropes, radios, harnesses, and masks.

Week 8 Day 1 (page 50)

1. Lake Winnipesaukee and Lake Champlain
2. Connecticut River
3. Answers should include two of the following: Passumpsic River, Watts River, White River, and West River.

ANSWER KEY *(cont.)*

Week 8 Day 3 (page 52)

1. Atlantic Ocean
2. Answers may include that chemicals can make people and animals sick or that watersheds can pollute rivers and oceans.
3. Answers may include that they are natural borders or that both states can use the resources from the river.

Week 8 Day 4 (page 53)

1. Answers may include there is too much trash and it is difficult to clean harmful liquids.
2. Answers may include that the river looks clean or the statistics show an improvement.
3. Answers should include a reason why the fact was surprising.

Week 9 Day 1 (page 55)

1. North. Answers should include that it is above the highway on the map.
2. Old Flat, Matthews, and Hawks
3. 23 state forests
4. Channels State Forest

Week 9 Day 2 (page 56)

Starting in the north and moving clockwise, the forests are: Coopers Rock, Kumbrabow, Seneca, Calvin Price, Greenbrier, Camp Creek, and Cabwaylingo.

Week 9 Day 3 (page 57)

1. Answers should include two of the following: hiking, biking, camping, fishing, and hunting.
2. Answers may include planting new trees, making sure the lake is stocked with fish, and conducting research.
3. "filled" or "full of"

Week 9 Day 4 (page 58)

1. 13 years; counting the rings
2. Answers may include lightning hit the tree, or people or animals damaged it.
3. Years 7 and 8; answers may include that the tree was damaged.

Week 10 Day 1 (page 60)

1. Fulton County
2. nine counties
3. northeast

Week 10 Day 2 (page 61)

Starting at Atlanta and moving clockwise—Fulton County, Richmond County, Chatham County, Houston County, and Bibb County.

Week 10 Day 3 (page 62)

1. Answers may include spending money on schools, choosing which roads to repair, and planning neighborhoods and parks.
2. over 900 square miles
3. Answers may include that it includes the capital city.

Week 10 Day 4 (page 63)

1. 546,503 people
2. Henry County, Forsyth County
3. Answers may include that there are more cities, more jobs, or better resources.

Week 11 Day 1 (page 65)

1. Wisconsin, Illinois, Kentucky, Tennessee, and Mississippi
2. Gulf of Mexico
3. Kentucky

Week 11 Day 3 (page 67)

1. fishing, drinking water, and traveling
2. They made travel much faster.
3. Answers may include that they helped transport goods such as cotton and other crops.

Week 11 Day 4 (page 68)

1. They are all on the Mississippi River.
2. Natchez Port
3. Answers may include Port Bienville, Gulfport, and Pascagoula since they are near the Gulf of Mexico.

Week 12 Day 1 (page 70)

1. southwest
2. west
3. northwest
4. Answers may include that it is in the southern, central part of the state.

Week 12 Day 2 (page 71)

Starting at Tallahassee (marked with a star) and moving clockwise: Tallahassee, Jacksonville, Orlando, Miami, and Tampa.

ANSWER KEY *(cont.)*

Week 12 Day 3 (page 72)

1. Answers may include alligators, snakes, birds, turtles, and lizards.
2. Answers may include that the blades of grass have sharp edges like a saw.
3. Answers should include that it is a meat-eating plant.

Week 12 Day 4 (page 73)

1. Answers should be between 600 and 700 pythons.
2. Answers may include that more people were looking for snakes or that more snakes were released.
3. Answers may include that fewer pythons were released into the Everglades or that more scientists helped capture them.

Week 13 Day 1 (page 75)

1. Annapolis
2. New Jersey
3. Westminster and La Plata
4. 40°N, 76°W
5. 39°N, 76°W

Week 13 Day 3 (page 77)

1. Ocean City and Dewey Beach
2. Answers should include two of the following: sailing, transporting goods, and carrying people.
3. Answers may include that there are many resources near the ocean, or people enjoy relaxing by the beach.

Week 13 Day 4 (page 78)

1. January, February, and March
2. The water is cold during the winter and rises through the summer. The water slowly cools into the winter months.
3. The water does not freeze because it never reaches the freezing temperature of 32°F.

Week 14 Day 1 (page 80)

1. reservoir
2. The force of the water running down the penstock spins the turbine.
3. The power lines carry the electricity to homes.

Week 14 Day 2 (page 81)

1. solar energy
2. biomass energy
3. wind energy
4. hydroelectric energy
5. geothermal energy

Week 14 Day 3 (page 82)

1. coal, oil, natural gas
2. A renewable resource is one that can be used over and over without running out.
3. Answers may include that the power plants could pollute the air or use up more of our nonrenewable resources.

Week 14 Day 4 (page 83)

1. Texas, Florida, Ohio
2. 6,430 billion cubic feet
3. six states
4. Answers may include that Texas is a large state or that Texas has more people.

Week 15 Day 1 (page 85)

1. about 100 miles
2. about 50 miles
3. No, the scale on the map shows that it is much shorter than 500 miles.

Week 15 Day 3 (page 87)

1. warm water
2. They are on the east coast, and hurricanes usually move northwest.
3. Students should circle the center of the hurricane.
4. Answers may include that it is very large because it is covering much of the Atlantic Ocean.

Week 15 Day 4 (page 88)

1. Answers should include damage to roofs and large branches will snap.
2. Answer may include that the winds don't get much stronger, or winds that strong are rare.

Week 16 Day 1 (page 90)

1. 60 inches
2. Texas and Oklahoma
3. 119 inches
4. 23 inches

ANSWER KEY *(cont.)*

Week 16 Day 2 (page 91)

1. Louisiana (purple); Texas (yellow); Missouri (green); Mississippi, Arkansas, Tennessee, Alabama (blue), and Oklahoma (orange)

Week 16 Day 3 (page 92)

1. A bayou is similar to a swamp with deeper, slow-moving water.
2. The land is unfit for humans to live on.
3. Answers may include getting lost or that wildlife can be dangerous.

Week 16 Day 4 (page 93)

1. The house is raised above the water.
2. Answers may include difficulty traveling, getting to stores, or dealing with wildlife.
3. Answers may include boating.

Week 17 Day 1 (page 95)

1. Alabama, Georgia, Tennessee, and Kentucky
2. Less than half. Explanations should reference the map.
3. Cumberland Mountains

Week 17 Day 2 (page 96)

clockwise from the top right: Allegheny Mountains, Cumberland Mountains, Cumberland Plateau, Allegheny Plateau

Week 17 Day 3 (page 97)

1. The sides of the plateau were hard to climb.
2. A ghost town is an abandoned town. The land ran out of coal and timber.
3. Answers may include swimming, hiking, exploring, or visiting state parks.

Week 17 Day 4 (page 98)

1. The two flat areas on either side of the mountain on the right should be labeled.
2. They are on both sides of the mountain on the right.
3. A valley dips low. A plateau is a long and flat area of land.
4. The water flows off the plateau's steep edge, which causes the waterfall.

Week 18 Day 1 (page 100)

1. Cook County
2. Answers should explain that the populations are high, but not as high as Cook County, or that the populations are between 300,000 and 949,999.
3. LaSalle
4. No, the population of Lake County is between 300,000–949,000.

Week 18 Day 2 (page 101)

starting at Rockford City (top) and moving clockwise: Rockford City, Aurora, Chicago, Bloomington, and Quincy City.

Week 18 Day 3 (page 102)

1. A metropolitan area is a city and all of its nearby towns and suburbs.
2. Answers may include that it is quieter, there are fewer people, and homes are more spread out.
3. Answers may include that places are close together, and people can live and work in tall skyscrapers.

Week 18 Day 4 (page 103)

1. Answers may include that there are many tall buildings close together.
2. Answers may include litter, water, air, light, or sound pollution.
3. Answers may include that trees have been cut down, and the land is covered by buildings and streets.
4. The tall building with two antennas should be circled.

Week 19 Day 1 (page 105)

1. alligator and tiger
2. eight exhibits
3. Example: *Head north past the fountain. Turn east to pass the alligator and flamingo. Turn north again at the cheetahs.*

Week 19 Day 2 (page 106)

ANSWER KEY *(cont.)*

Week 19 Day 3 (page 107)

1. A tourist is a person who visits a place to see the sights.
2. Answers may include Rock and Roll Hall of Fame, Cedar Point, King's Island, and the Pro Football Hall of Fame.
3. Answers may include that it is exciting to watch.

Week 19 Day 4 (page 108)

1. Answers may include preparing food, cleaning, and mowing the grass.
2. Answers may include selling food and souvenirs and providing security.
3. Answers may include picking up trash and helping people exit.

Week 20 Day 1 (page 110)

1. Three Affiliated Tribes Museum and Akta Lakota Museum
2. North Dakota: Bismarck; South Dakota: Pierre; answers should mention using the legend.
3. Answers should describe that the trail goes through the center of the states, curving west through North Dakota and east through South Dakota.

Week 20 Day 3 (page 112)

1. Thomas Jefferson, Abraham Lincoln, George Washington, and Theodore Roosevelt
2. Answers may include that it is a symbol of freedom or it shows four great presidents.
3. Answers may include that it is unique or that it shows U.S. history.

Week 20 Day 4 (page 113)

1. Sites 5 and 10; they are marked with a wheelchair symbol.
2. 13 spigots
3. Little Missouri River; answers may include swimming, fishing, and boating.

Week 21 Day 1 (page 115)

1. Answers may include that some of the land is worn away.
2. Mountains became lower, and rivers sunk deeper.
3. Answers should be supported with details from the images.

Week 21 Day 3 (page 117)

1. A windbreak is a line of trees or bushes that block the wind.
2. The roots help hold the soil together.
3. Answers may include planting grass, using less water, and planting a windbreak.

Week 21 Day 4 (page 118)

1. Upper Mississippi, Lower Mississippi, Arkansas Red-White, and Missouri
2. Answers may include that Missouri is part of many river basins.

Week 22 Day 1 (page 120)

1. cropland
2. They are both urban areas.
3. cattle, pigs, chicken, and sheep

Week 22 Day 3 (page 122)

1. Answers may include that the land is divided into sections to farm in rural areas, and that there are buildings, roads, and many people who live in urban areas.
2. Answers may include that the city was changed by building roads and tall buildings, while the farmland was changed by dividing it into sections to farm.

Week 22 Day 4 (page 123)

1. Eastern side of the United States; answers should reference the map.
2. About 10 million people; answers should include adding the populations shown as bars.
3. Answers may include that most of the map would not contain as many bars.

Week 23 Day 1 (page 125)

1. The field is mostly in squares E1, F1, E2, F2, E3, and F3.
2. The sheep are in squares A3, B3, A4, B4, and B5.
3. a tree
4. cows
5. farmhouse

Week 23 Day 3 (page 127)

1. corn, soybeans, and apples
2. A plow loosens the soil for planting.
3. Answers may include that the tools make it easier and cheaper to farm more land.

ANSWER KEY *(cont.)*

Week 23 Day 4 (page 128)

1. five products
2. No, the combined amount is 156,499,000 lbs.
3. Answers should describe that it is a top dairy producer in the United States.
4. Answers may include that it produces a lot of dairy.

Week 24 Day 1 (page 130)

1. Lake Michigan and Lake Superior
2. Huron, Ontario, Michigan, Erie, Superior

Week 24 Day 2 (page 131)

From north to south: Lake Superior, Lake Huron, Lake Michigan, Lake Ontario, and Lake Erie

Week 24 Day 3 (page 132)

1. Answers may include that the lakes are very large.
2. Answers may include fishing, boating, or swimming.
3. Similarities may include that it is very large and goods are shipped on it. Differences may include that the lake has freshwater.

Week 24 Day 4 (page 133)

1. True, Indiana has about 50,000 jobs; Ohio has about 175,000.
2. about 550,000 jobs
3. Answers may include that Michigan borders four of the five lakes.

Week 25 Day 1 (page 135)

1. Pushmataha County
2. storm watch or stormy weather
3. central and eastern part of the state
4. Warning; it is shown lower and darker in the legend.

Week 25 Day 3 (page 137)

1. the National Weather Service
2. Tornado Alley is a region where many tornados occur.
3. A tornado warning should be issued since a tornado has formed.

Week 25 Day 4 (page 138)

1. EF-0
2. An EF-5 causes major destruction. Homes and trees are leveled.
3. Answers may include that blowing debris or falling materials are dangerous.

Week 26 Day 1 (page 140)

1. Mojave Desert
2. Chihuahuan Desert
3. Answers may include that the climate will be hot and dry.

Week 26 Day 2 (page 141)

From top to bottom: Great Basin Desert, Mojave Desert, Sonoran Desert, and Chihuahua Desert.

Week 26 Day 3 (page 142)

1. They have sharp needles or leaves, and they can store water.
2. They burrow in the soil or hide in shady areas.
3. Answers may include that they can hide from enemies, and stay out of the heat.

Week 26 Day 4 (page 143)

1. The cliff provides shade for the homes.
2. Answers may include rocks or soil.
3. Answers may include the extreme temperatures and the high location of the homes.
4. Answers may include that the homes are shaded and they are protected by the cliff.

Week 27 Day 1 (page 145)

1. three missions
2. eight missions
3. Answers may include that the Spanish controlled the land for many years.

Week 27 Day 3 (page 147)

1. Answers may include that Texan soldiers chose to fight knowing they would lose.
2. to protect Mexico from its enemies
3. Answers may include that it has an inspiring story and is part of history.

ANSWER KEY *(cont.)*

Week 27 Day 4 (page 148)

1. Answers may include that they are part of Texas history.
2. Answers may include that they brought soil from other areas.
3. Answers may include that rain and wind might erode the mounds and land.

Week 28 Day 1 (page 150)

1. Mormon Trail
2. Mormon Trail and Oregon Trail
3. They would follow the Oregon Trail until Idaho where they would then take the California trail.

Week 28 Day 2 (page 151)

From top to bottom: Oregon Trail, California Trail, Mormon Trail, and Santa Fe Trail.

Week 28 Day 3 (page 152)

1. mountains, plateaus, and canyons
2. near the Great Salt Lake
3. Answers may include a hot climate, little water, and dangerous animals.

Week 28 Day 4 (page 153)

1. about 600 miles
2. There were no miles of track in 1840.
3. Answers may include that as people began using cars and trucks, trains weren't used as much.

Week 29 Day 1 (page 155)

1. five national forests
2. 11 national forests
3. Rogue River-Siskiyou National Forest
4. Umatilla National Forest

Week 29 Day 2 (page 156)

Starting in the northeast corner and going clockwise: Colville, Umatilla, Gifford Pinchot, Olympic, Mt. Baker-Snoqualmie, and Okanogan and Wenatchee.

Week 29 Day 3 (page 157)

1. Reforestation is the process of replanting trees that have been cut down.
2. Answers may include tools, boxes, and sports equipment.
3. Deforestation, since the trees have been cut down.

Week 29 Day 4 (page 158)

1. 2014
2. Answers may include that there was a lot of lightning in some years, or the land was too dry.
3. 18,822 acres
4. Answers may include reporting small fires, putting out campfires, and being safe around fires.

Week 30 Day 1 (page 160)

1. Colorado
2. Rocky Mountains
3. New Mexico, Wyoming, Idaho, Utah, Montana
4. 40°N, 105°W

Week 30 Day 2 (page 161)

From top to bottom: Granite Peak, Francis Peak, Mount Elbert, and Culebra Peak.

Week 30 Day 3 (page 162)

1. Rocky Mountain National Park
2. Answers may include skiing, snowboarding, hiking, climbing, camping, driving, fishing, horseback riding, and going on tours.
3. 14,443 ft. (4,402 m)

Week 30 Day 4 (page 163)

1. Teakettle, Whitehouse, Mt. Ridgeway, Corbett, and Cirque
2. Answers may include that the peaks are all very similar in height and are all between 13,000 and 14,000 feet.
3. Whitehouse and Mt. Ridgeway
4. Answers may include that a person would want to ski or snowboard, see the view from up high, and see wildlife.

Week 31 Day 1 (page 165)

1. Canada
2. China, Singapore, Japan, and Germany
3. airplanes and ships

Week 31 Day 2 (page 166)

Starting in the center and moving clockwise: United Kingdom, Germany, China, Japan, Singapore, Mexico, and Canada.

ANSWER KEY *(cont.)*

Week 31 Day 3 (page 167)

1. wood and food
2. metals and farming products
3. paper products, clothes, and leather
4. potatoes

Week 31 Day 4 (page 168)

1. South Korea
2. Canada
3. about 24 percent
4. Answers may include that Canada is near Idaho.

Week 32 Day 1 (page 170)

1. 10 states
2. North Dakota, South Dakota, Nebraska, Kansas
3. Rocky Mountains

Week 32 Day 3 (page 172)

1. Answers should include hot summers, cold winters, and little rainfall.
2. Answers may include that unlike the rest of the plains, it is very tall.
3. Example: *I would rather visit the eastern portion because then I could see Devil's Tower.*

Week 32 Day 4 (page 173)

1. West Yellowstone; the temperatures are lower.
2. 17°F
3. July

Week 33 Day 1 (page 175)

1. exceptional drought; it is last in the scale.
2. Reno, Carson City
3. abnormally dry
4. only the northeast corner of the state

Week 33 Day 3 (page 177)

1. Answers may include that plants and animals can become dehydrated, water parks may close, fountains may be turned off, and people may need to buy water.
2. Answers should include two of the following: water lawns early or late, use less water to wash cars, take shorter showers, or run dishwashers or washing machines only when full.
3. Answers may include dry and cracked soil or the dead tree.

Week 33 Day 4 (page 178)

1. four months
2. The Spring months of March, April, May receive the most rain.
3. June and July, since these are the typically the driest months.

Week 34 Day 1 (page 180)

1. four fault lines
2. Oakland
3. San Andreas Fault
4. Newport-Inglewood Fault

Week 34 Day 2 (page 181)

From top to bottom: Eureka, Oakland, San Francisco, Fresno, Los Angeles, and San Diego,

Week 34 Day 3 (page 182)

1. near San Francisco
2. Answers should include two of the following: buildings crumbled, streets cracked, water pipes burst, fires started, people died, people were left homeless, and the city was in ruins.
3. Answers may include that scientists are studying the fault, new technologies can better predict earthquakes, and buildings are made to withstand earthquakes.
4. The fault crack in the ground should be circled.

Week 34 Day 4 (page 183)

1. Answers should be supported by logical reasons.
2. Answers may include that it could be several days before they receive help.
3. Answers may include bandages, cotton balls, and medicine.

Week 35 Day 1 (page 185)

1. beneath the Earth's surface or below the volcano
2. crater
3. Answers may include that it is wide with a gentle slope.
4. Answers may include that lava spreads out over the side of the volcano.

Week 35 Day 2 (page 186)

The crater is is the hole in the center of the photo. The lava is the bright spot in the middle of the crater. Descriptions should include seeing a crater and lava.

ANSWER KEY *(cont.)*

Week 35 Day 3 (page 187)

1. Answers may include that the lava is very hot.
2. It is low and flat, like a shield.
3. Answers may include that there is steam and the lava is glowing.

Week 35 Day 4 (page 188)

1. the island of Hawai'i
2. Answers may include that the lava flows slowly, and people keep their distance.
3. Answers may include that it is a unique thing to see.

Week 36 Day 1 (page 190)

1. Juneau
2. Arctic Ocean, Pacific Ocean, and Bering Sea
3. Russia
4. The tundra is in the northern and western part of the state.

Week 36 Day 3 (page 192)

1. They receive little rainfall, and it's very cold.
2. Much of the soil is frozen. The growing season is too short.
3. the Arctic Ocean

Week 36 Day 4 (page 193)

1. Answers may include there are few trees or plants, and the soil is covered with snow.
2. Answers may include warm shelters and clothing, and hunting and fishing rather than farming.
3. Example answer: *No, because crops cannot grow due to frozen soil and little rain.*

POLITICAL MAP OF THE UNITED STATES

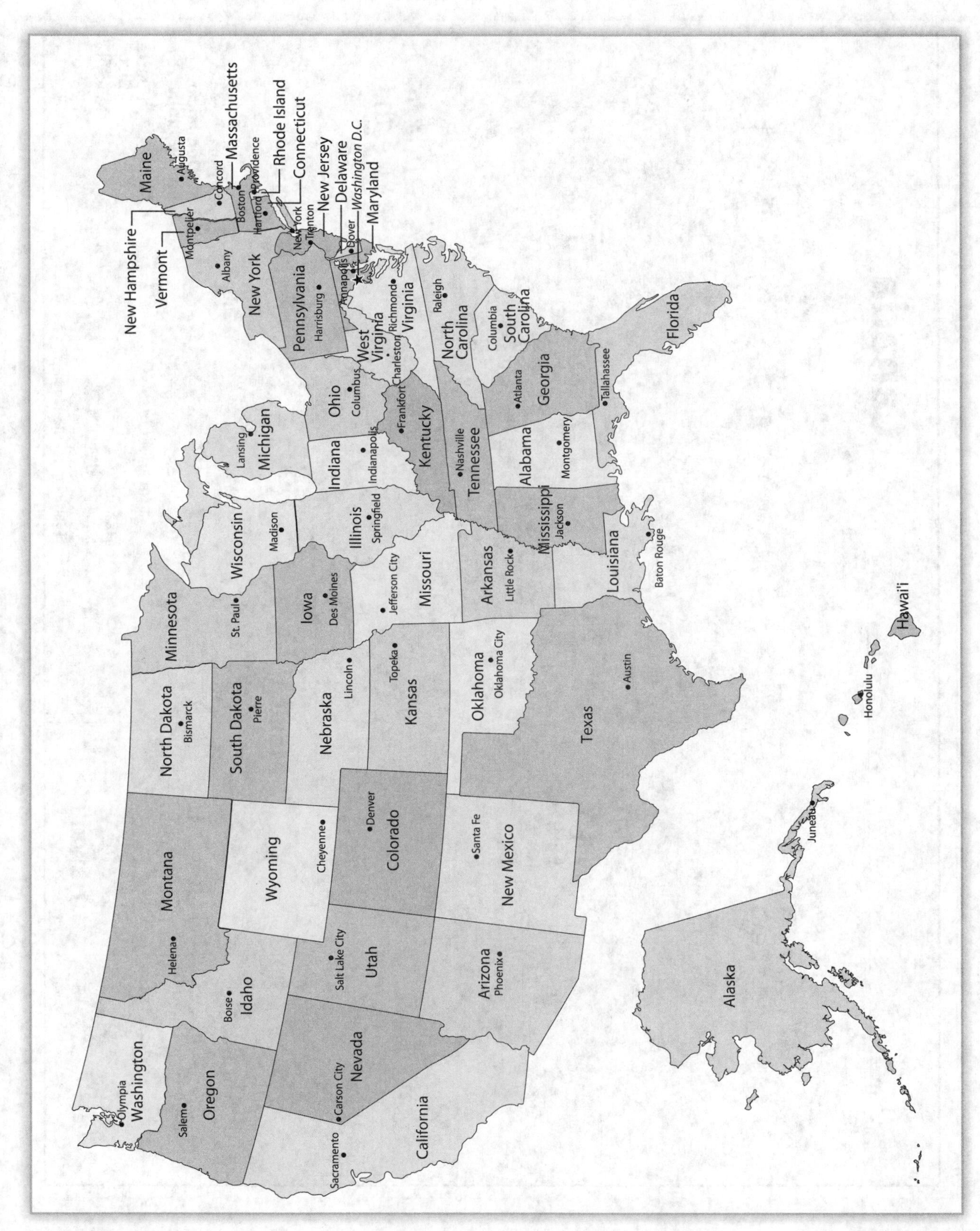

PHYSICAL MAP OF THE UNITED STATES

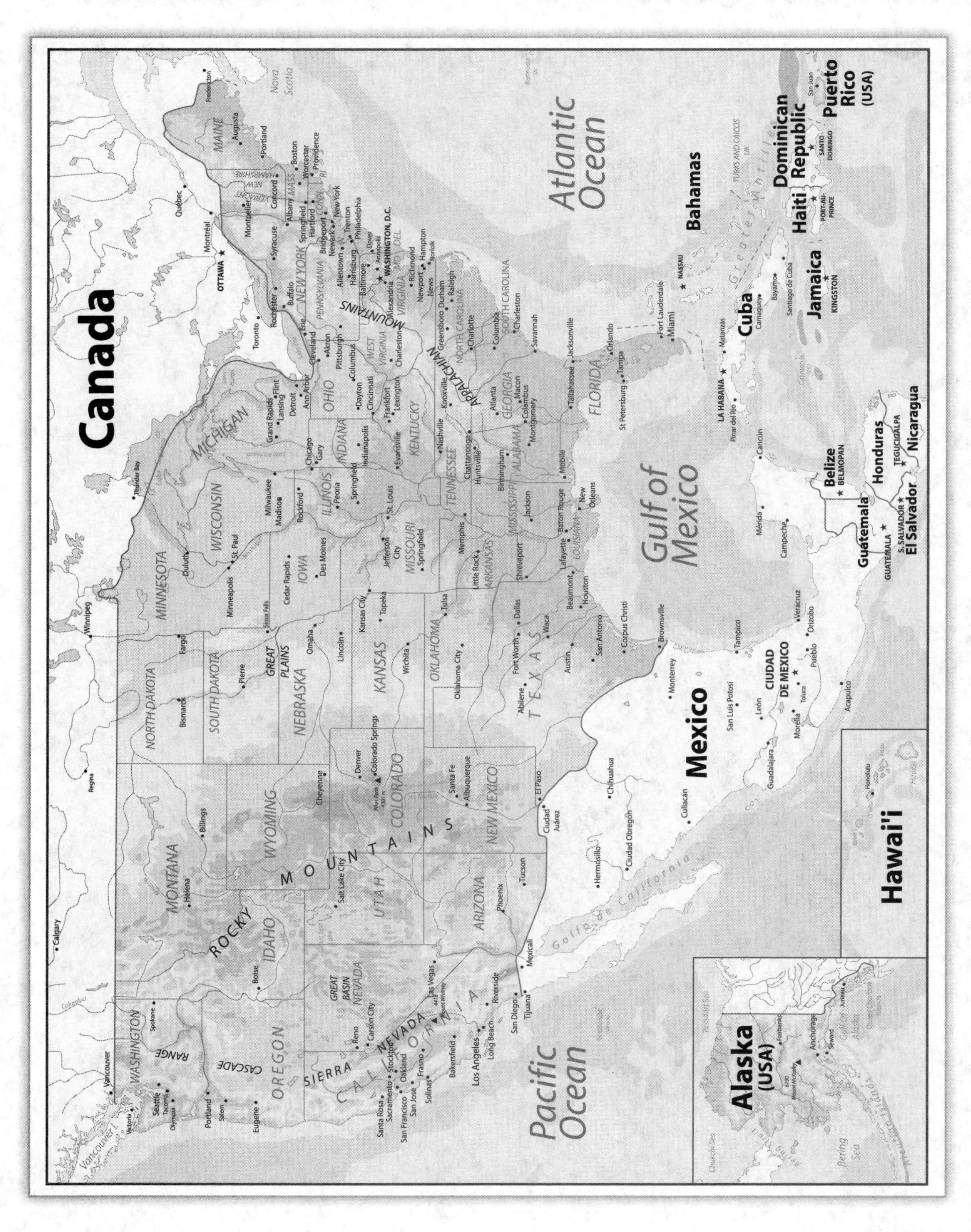

WESTERN HEMISPHERE

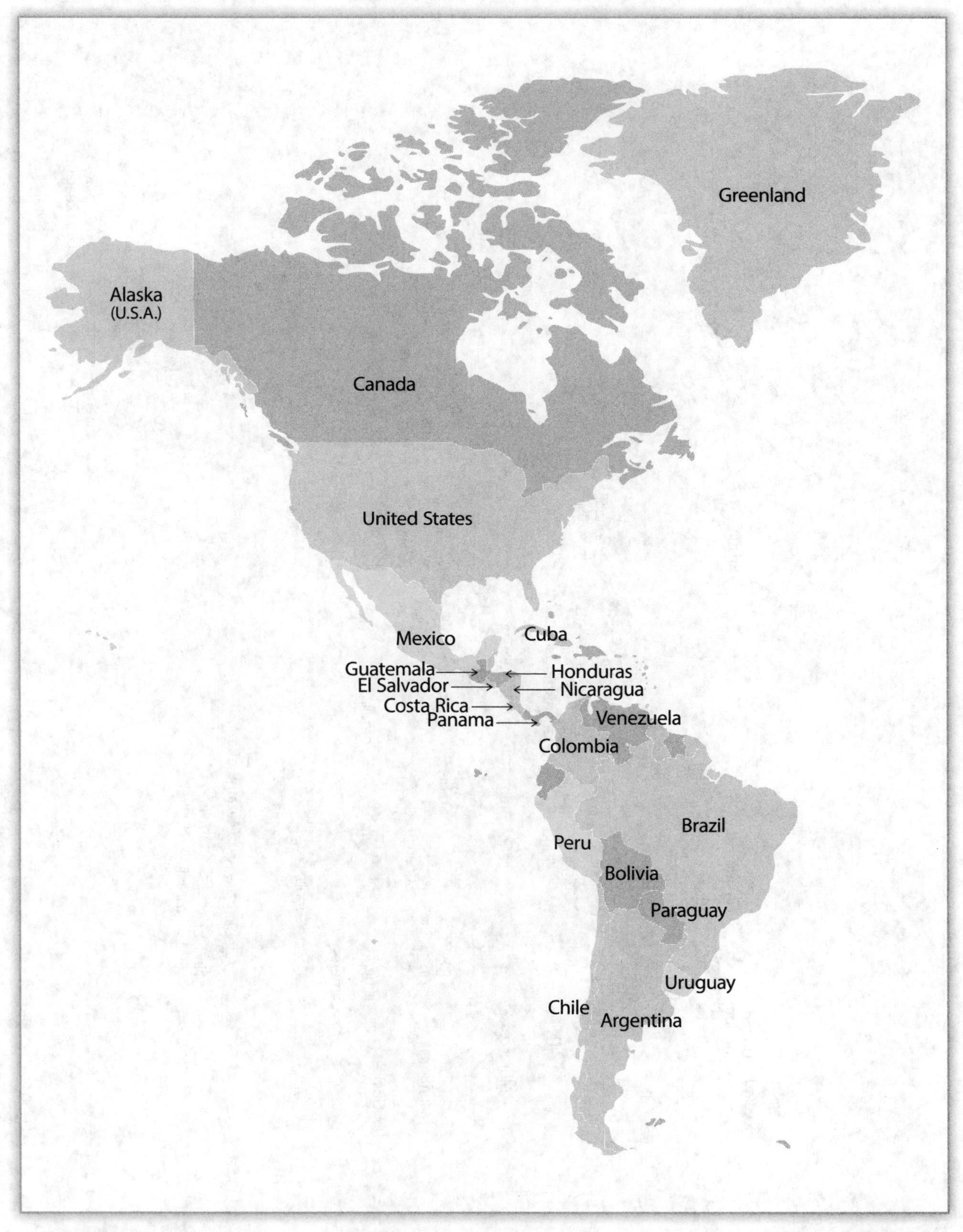

EASTERN HEMISPHERE

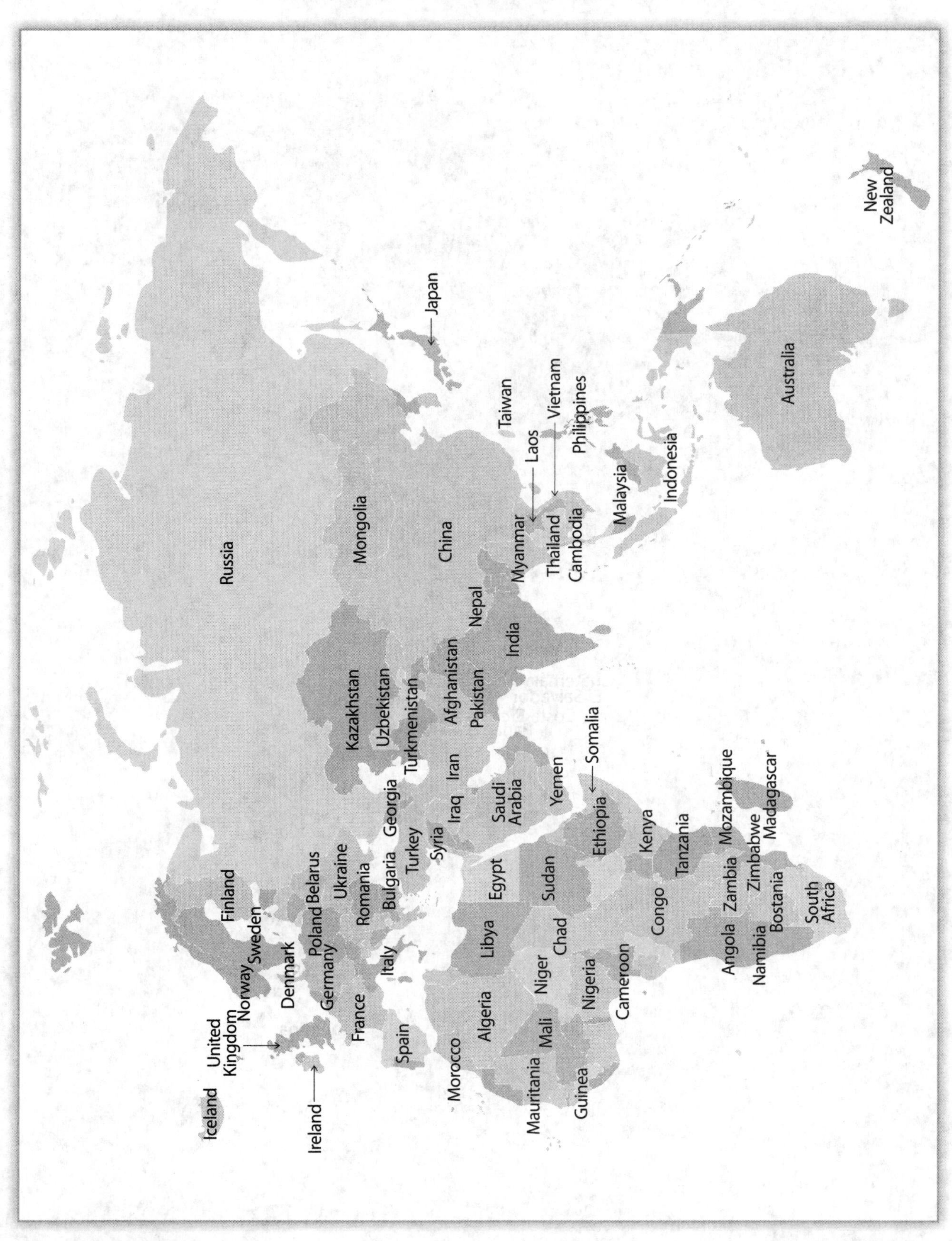

WORLD

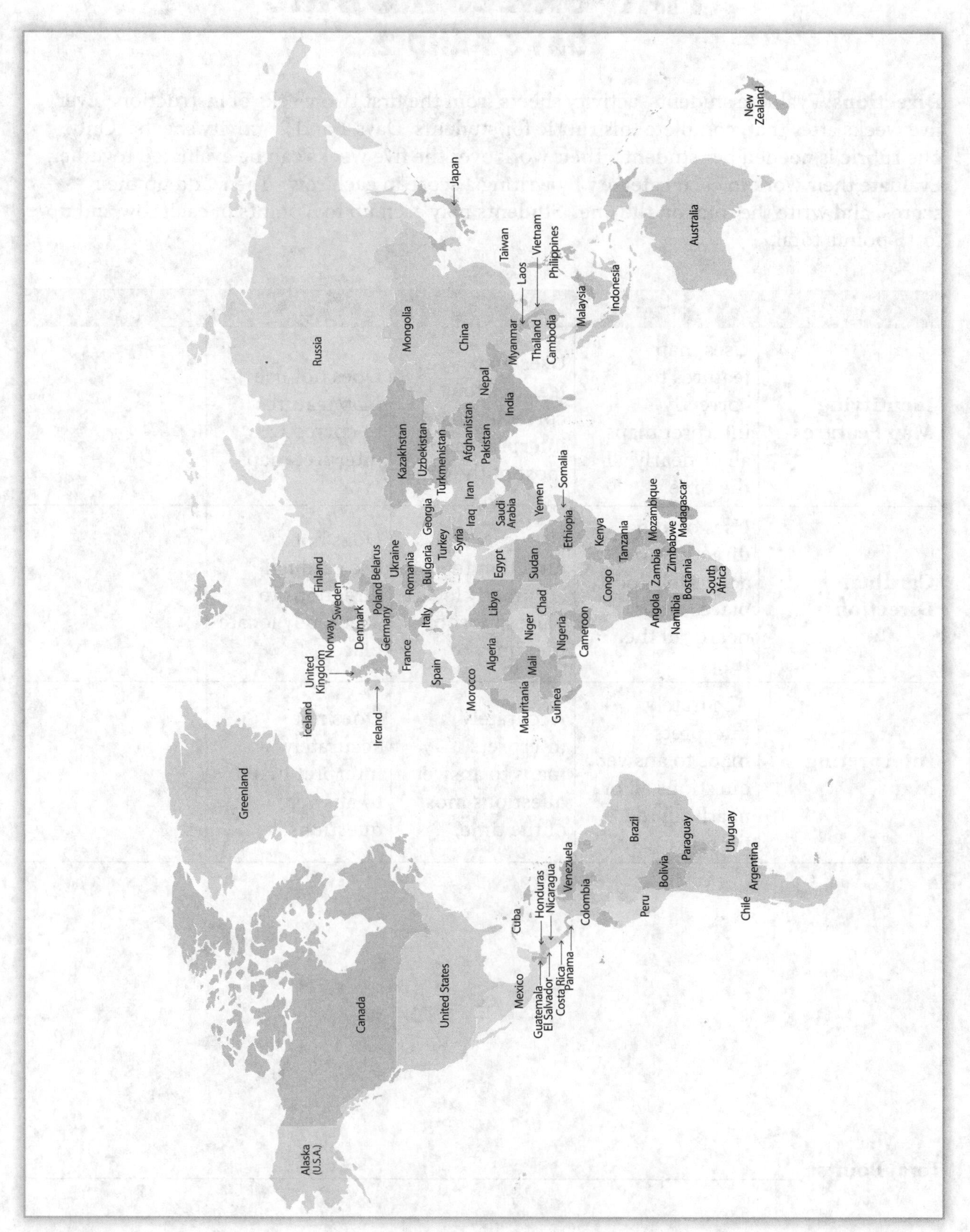

Alaska (U.S.A.)
Canada
United States
Mexico
Guatemala
El Salvador
Costa Rica
Panama
Honduras
Nicaragua
Cuba
Venezuela
Colombia
Peru
Brazil
Bolivia
Paraguay
Uruguay
Chile
Argentina
Greenland
Iceland
Ireland
United Kingdom
Norway
Sweden
Finland
Denmark
Germany
Poland
Belarus
Ukraine
France
Spain
Italy
Romania
Bulgaria
Turkey
Georgia
Syria
Iraq
Iran
Saudi Arabia
Yemen
Morocco
Algeria
Libya
Egypt
Sudan
Chad
Niger
Mali
Mauritania
Guinea
Nigeria
Cameroon
Ethiopia
Somalia
Kenya
Congo
Tanzania
Angola
Zambia
Namibia
Bostania
Zimbabwe
Mozambique
Madagascar
South Africa
Russia
Kazakhstan
Uzbekistan
Turkmenistan
Afghanistan
Pakistan
India
Nepal
Mongolia
China
Myanmar
Thailand
Cambodia
Laos
Vietnam
Taiwan
Philippines
Malaysia
Indonesia
Japan
Australia
New Zealand

Name: ______________________ Date: ______________

MAP SKILLS RUBRIC
DAYS 1 AND 2

Directions: Evaluate students' activity sheets from the first two weeks of instruction. Every five weeks after that, complete this rubric for students' Days 1 and 2 activity sheets. Only one rubric is needed per student. Their work over the five weeks can be evaluated together. Evaluate their work in each category by writing a score in each row. Then, add up their scores, and write the total on the line. Students may earn up to 5 points in each row and up to 15 points total.

Skill	5	3	1	Score
Identifying Map Features	Uses map features to correctly interpret maps all or nearly all the time.	Uses map features to correctly interpret maps most of the time.	Does not use map features to correctly interpret maps.	
Cardinal Directions	Uses cardinal direction to accurately locate places all or nearly all the time.	Uses cardinal direction to accurately locate places most of the time.	Does not use cardinal directions to accurately locate places.	
Interpreting Maps	Accurately interprets maps to answer questions all or nearly all the time.	Accurately interprets maps to answer questions most of the time.	Does not accurately interpret maps to answer questions.	

Total Points: ______________________

Name: ______________________________ Date: ____________________

APPLYING INFORMATION AND DATA RUBRIC

DAYS 3 AND 4

Directions: Complete this rubric every five weeks to evaluate students' Day 3 and Day 4 activity sheets. Only one rubric is needed per student. Their work over the five weeks can be evaluated together. Evaluate their work in each category by writing a score in each row. Then, add up their scores, and write the total on the line. Students may earn up to 5 points in each row and up to 15 points total. **Note:** Weeks 1 and 2 are map skills only and will not be evaluated here.

Skill	5	3	1	Score
Interpreting Text	Correctly interprets texts to answer questions all or nearly all the time.	Correctly interprets texts to answer questions most of the time.	Does not correctly interpret texts to answer questions.	
Interpreting Data	Correctly interprets data to answer questions all or nearly all the time.	Correctly interprets data to answer questions most of the time.	Does not correctly interpret data to answer questions.	
Applying Information	Applies new information and data to known information about locations or regions all or nearly all the time.	Applies new information and data to known information about locations or regions most of the time.	Does not apply new information and data to known information about locations or regions.	

Total Points: ______________________________

Name: ______________________________ **Date:** ____________________

MAKING CONNECTIONS RUBRIC
DAY 5

Directions: Complete this rubric every five weeks to evaluate students' Day 5 activity sheets. Only one rubric is needed per student. Their work over the five weeks can be evaluated together. Evaluate their work in each category by writing a score in each row. Then, add up their scores, and write the total on the line. Students may earn up to 5 points in each row and up to 15 points total. **Note:** Weeks 1 and 2 are map skills only and will not be evaluated here.

Skill	5	3	1	Score
Comparing One's Community	Makes meaningful comparisons of one's own home or community to others all or nearly all the time.	Makes meaningful comparisons of one's own home or community to others most of the time.	Does not make meaningful comparisons of one's own home or community to others.	
Comparing One's Life	Makes meaningful comparisons of one's daily life to those in other locations or regions all or nearly all the time.	Makes meaningful comparisons of one's daily life to those in other locations or regions most of the time.	Does not makes meaningful comparisons of one's daily life to those in other locations or regions.	
Making Connections	Uses information about other locations or regions to make meaningful connections about life there all or nearly all the time.	Uses information about locations or regions to make meaningful connections about life there most of the time.	Does not use information about locations or regions to make meaningful connections about life there.	

Total Points: ____

MAP SKILLS ANALYSIS

Directions: Record each student's rubric scores (page 210) in the appropriate columns. Add the totals, and record the sums in the Total Scores column. Record the average class score in the last row. You can view: (1) which students are not understanding map skills and (2) how students progress throughout the school year.

Student Name	Week 2	Week 7	Week 12	Week 17	Week 22	Week 27	Week 32	Week 36	Total Scores
Average Classroom Score									

APPLYING INFORMATION AND DATA ANALYSIS

Directions: Record each student's rubric scores (page 211) in the appropriate columns. Add the totals, and record the sums in the Total Scores column. Record the average class score in the last row. You can view: (1) which students are not understanding how to analyze information and data and (2) how students progress throughout the school year.

Student Name	Week 7	Week 12	Week 17	Week 22	Week 27	Week 32	Week 36	Total Scores
Average Classroom Score								

MAKING CONNECTIONS ANALYSIS

Directions: Record each student's rubric scores (page 212) in the appropriate columns. Add the totals, and record the sums in the Total Scores column. Record the average class score in the last row. You can view: (1) which students are not understanding how to make connections to geography and (2) how students progress throughout the school year.

Student Name	Week 7	Week 12	Week 17	Week 22	Week 27	Week 32	Week 36	Total Scores
Average Classroom Score								

DIGITAL RESOURCES

To access the digital resources, go to this website and enter the following code: 71184716.
www.teachercreatedmaterials.com/administrators/download-files/

Rubrics

Resource	Filename
Map Skills Rubric	skillsrubric.pdf
Applying Information and Data Rubric	datarubric.pdf
Making Connections Rubric	connectrubric.pdf

Item Analysis Sheets

Resource	Filename
Map Skills Analysis	skillsanalysis.pdf skillsanalysis.docx skillsanalysis.xlsx
Applying Information and Data Analysis	dataanalysis.pdf dataanalysis.docx dataanalysis.xlsx
Making Connections Analysis	connectanalysis.pdf connectanalysis.docx connectanalysis.xlsx

Standards

Resource	Filename
Standards Charts	standards.pdf